INSTITUTE OF GEOLOGICAL SCIENCES

Natural Environment Research Council

8000045378
TREMOUGH CAMPUS LIBRARY

AF412748

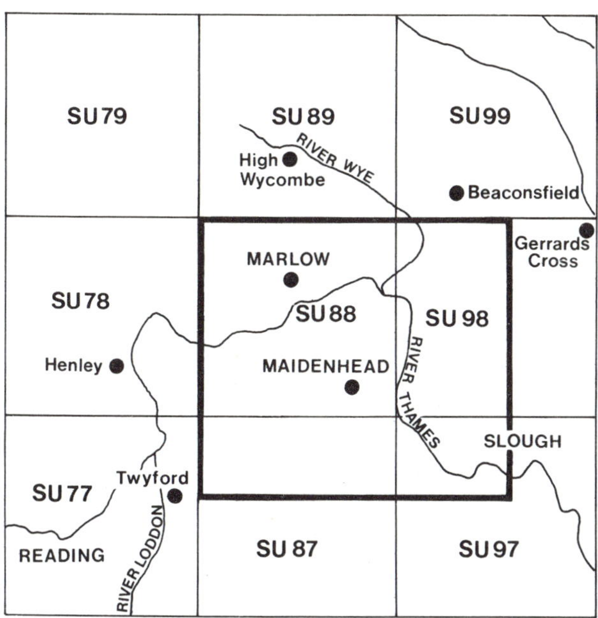

The sand and gravel resources of the country around Maidenhead and Marlow

Description of 1:25 000 resource sheet SU 88 and parts SU 87, 97, and 98

P. N. Dunkley

Tremough Campus Library
STANDARD LOAN
This book may be borrowed for 28 days, subject to recall

Please return or renew this book on or before the due date, after which daily fines will be
added to your account. Please check your issue slip or log onto 'my account' at
http//voyager.falmouth.ac.uk

Library & Information Services
Tremough Campus, Penryn, Cornwall TR10 9EZ
01326 370441

http//library.falmouth.ac.uk or email library@falmouth.ac.uk

© Crown copyright 1979

ISBN 0 11 884091 6

London   Her Majesty's Stationery Office   1979

The first twelve reports on the assessment of British
sand and gravel resources appeared in the Report series
of the Institute of Geological Sciences as a subseries.
Report 13 and subsequent reports appear as
Mineral Assessment Reports of the Institute.

Details of published reports appear at the end of this
Report.

Any enquiries concerning this report may be addressed
to Head, Industrial Minerals Assessment Unit,
Institute of Geological Sciences, Keyworth, Nottingham
NG12 5GG.

The asterisk on the cover indicates that parts of sheets
adjacent to the one cited are described in this report.

PREFACE

National resources of many industrial minerals may
seem so large that stocktaking appears unnecessary, but
the demand for minerals and for land for all purposes is
intensifying and it has become increasingly clear in
recent years that regional assessments of the resources
of these minerals should be undertaken. The
publication of information about the quantity and
quality of deposits over large areas is intended to
provide a comprehensive factual background against
which planning decisions can be made.

Sand and gravel, considered together as naturally
occurring aggregate was selected as the bulk mineral
demanding the most urgent attention, initially in the
south-east of England, where about half the national
output is won and very few sources of alternative
aggregates are available. Following a short feasibility
project, initiated in 1966 by the Ministry of Land and
Natural Resources, the Industrial Minerals Assessment
Unit (formerly the Mineral Assessment Unit) began
systematic surveys in 1968. The work is now being
financed by the Department of the Environment and is
being undertaken with the cooperation of the Sand and
Gravel Association of Great Britain.

This report describes the sand and gravel resources
of the country around Maidenhead and Marlow, shown
on the accompanying 1:25 000 resource map which
comprises Ordnance Survey sheet SU 88 and parts of
SU 87, 97 and 98. The survey was conducted by
Dr H. C. Squirrell, assisted by Mr C. E. Corser, who
supervised the drilling and sampling programme.
Dr P. N. Dunkley compiled the report. The work was
based on a geological survey at 1:10 560 in 1902–1920
by R. L. Sherlock, A. H. Noble, C. N. Bromehead,
J. H. Blake and G. Barrow, with amendments by
H. C. Squirrell in 1971. The late Mr A. P. Mace (Land
Agent) was responsible for negotiating access to the
land for drilling; the ready cooperation of the land
owners and tenants in this work is much appreciated.
Information provided by local gravel operators and the
Berkshire and Buckinghamshire county councils is
gratefully acknowledged.

Austin W. Woodland
Director

Institute of Geological Sciences
Exhibition Road
London SW7 2DE

25 October 1978

CONTENTS

The sand and gravel resources of the country around Maidenhead and Marlow

Description of 1:25 000 resource sheet SU 88 and parts SU 87, 97 and 98

P. N. DUNKLEY

SUMMARY

The geological maps of the Institute of Geological Sciences, pre-existing borehole information, and 61 boreholes drilled for the Industrial Minerals Assessment Unit form the basis of the assessment of sand and gravel resources in the Maidenhead–Marlow area.

All deposits in the area which might be potentially workable for sand and gravel have been investigated and a simple statistical method has been employed to estimate the volume. The reliability of the volume estimates is given at the symmetrical 95 per cent probability level.

The 1:25 000 map is divided into six resource blocks containing between 9.1 and 11.7 km² of potentially workable sand and gravel. The geology of the deposits is described and the mineral-bearing area, the mean thickness of the overburden and mineral, and the mean grading of the mineral are stated. Detailed borehole data are given. The geology, the position of the boreholes and the outlines of the resource blocks are shown on the accompanying map.

Bibliographical reference

DUNKLEY, P. N. 1979. The sand and gravel resources of the country around Maidenhead and Marlow: description of 1:25 000 resource sheet SU 88 and parts SU 87, 97 and 98. *Miner. Assess. Rep. Inst. Geol. Sci.*, No. 42.

Author
P. N. Dunkley, BSc, PhD
Institute of Geological Sciences
Keyworth, Nottingham NG12 5GG

INTRODUCTION

The survey is concerned with the estimation of resources, which include deposits that are not currently exploitable but have a foreseeable use, rather than reserves, which can only be assessed in the light of current, locally prevailing, economic considerations. Clearly, both the economic and the social factors used to decide whether a deposit may be workable in the future cannot be predicted; they are likely to change with time. Deposits not currently economically workable may be exploited as demand increases, as higher grade or alternative materials become scarce, or as improved processing techniques are applied to them. The improved knowledge of the main physical properties of the resource and their variability which this survey seeks to provide, will add significantly to the factual background against which planning policies can be decided (Archer, 1969; Thurrell, 1971; Harris and others, 1974).

The survey provides information at the 'indicated' level 'for which tonnage and grade are computed partly from specific measurements, samples or production data and partly from projection for a reasonable distance on geological evidence. The sites available for inspection, measurement, and sampling are too widely spaced to permit the mineral bodies to be outlined completely or the grade established throughout'. (Bureau of Mines and Geological Survey, 1948, p. 15).

It follows that the whereabouts of reserves must still be established and their size and quality proved by the customary detailed exploration and evaluation undertaken by the industry. However, the information provided by this survey should assist in the selection of the best targets for such further work. The following arbitrary physical criteria have been adopted:

a The deposit should average at least one metre in thickness.
b The ratio of overburden to sand and gravel should be no more than 3:1.
c The proportion of fines (particles passing the No. 240 mesh BS sieve, about $\frac{1}{16}$ mm) should not exceed 40 per cent.
d The deposit must lie within 25 m of the surface, this being taken as the likely maximum working depth under most circumstances. It follows from the second criterion that boreholes are drilled no deeper than 18 m if no sand and gravel has been proved.

A deposit of sand and gravel which broadly meets these criteria is regarded as 'potentially workable' and is described and assessed as 'mineral' in this report. As the assessment is at the indicated level, parts of such a deposit may not satisfy all the criteria.

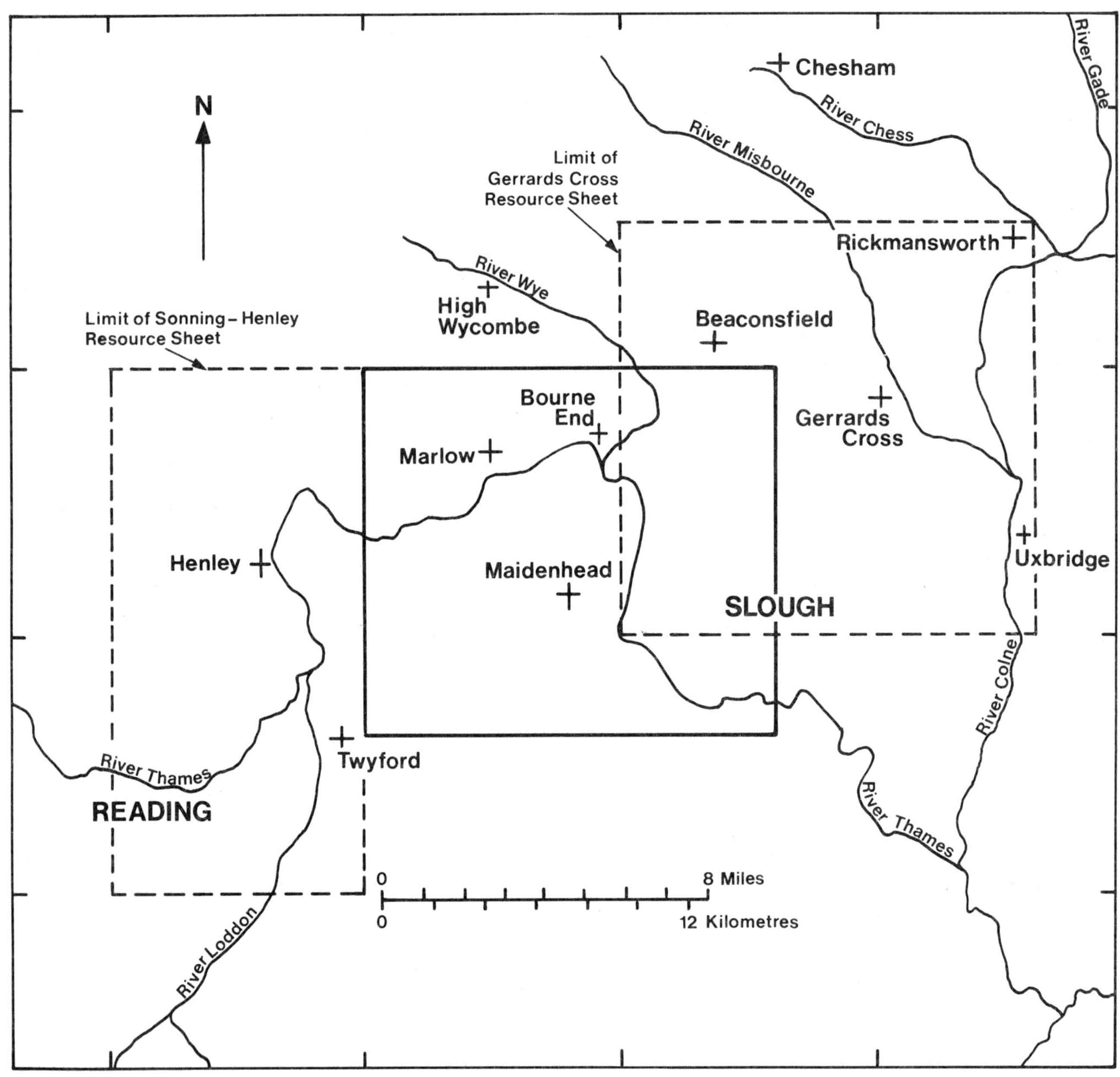

Figure 1   Sketch-map showing the location of the resource sheet

For the particular needs of assessing sand and gravel resources, a grain-size classification based on the geometric scale $\frac{1}{16}$ mm, $\frac{1}{4}$ mm, 1 mm, 4 mm, 16 mm has been adopted. The boundaries between fines (that is, the clay and silt fractions) and sand, and between sand and gravel grade material, are placed at $\frac{1}{16}$ mm and 4 mm respectively (see Appendix C).

The volume and other characteristics are assessed within resource blocks, each of which, ideally, contains approximately 10 km² of sand and gravel. No account is taken of any factors, for example, roads, villages and high agricultural or landscape value, which might stand in the way of sand and gravel being exploited, although towns are excluded. The estimated total volume therefore bears no simple relationship to the amount that could be extracted in practice.

It must be emphasised that the assessment applies to the resource block as a whole. Valid conclusions cannot be drawn about the mineral in parts of a block, except in the immediate vicinity of the actual sample points.

DESCRIPTION OF THE RESOURCE SHEET

GENERAL

The resource sheet covers a total area of 224 km² and comprises 1:25 000 Ordnance Survey sheet SU 88 and parts of SU 87, 97 and 98 (Figure 1). The area assessed amounts to 92.2 km² of which 62.5 km² is mineral bearing. No assessment has been made of the sand and gravel deposits covering 23.9 km² within the built-up areas of Slough, Maidenhead, Marlow, Windsor, Bourne End and Wooburn Green (Figure 2). In the north-east 33.9 km² of Glacial Sand and Gravel has been assessed in a previous report (Squirrell, 1974).

The sand and gravel deposits fall into two main categories, the river deposits and the glacial deposits. The river deposits (blocks A to E) cover an area of 51.9 km² within the valleys of the Thames and Wye, and the glacial deposits (Block F) occupy an area of 10.6 km² on higher ground above these valleys. The Thames Valley has been an important source of aggregate in the past, and from worked-out areas shown on the resource map an estimated 19 million m³ of sand and gravel has been extracted. The glacial deposits are more variable in thickness and grading, with a higher fines content than the

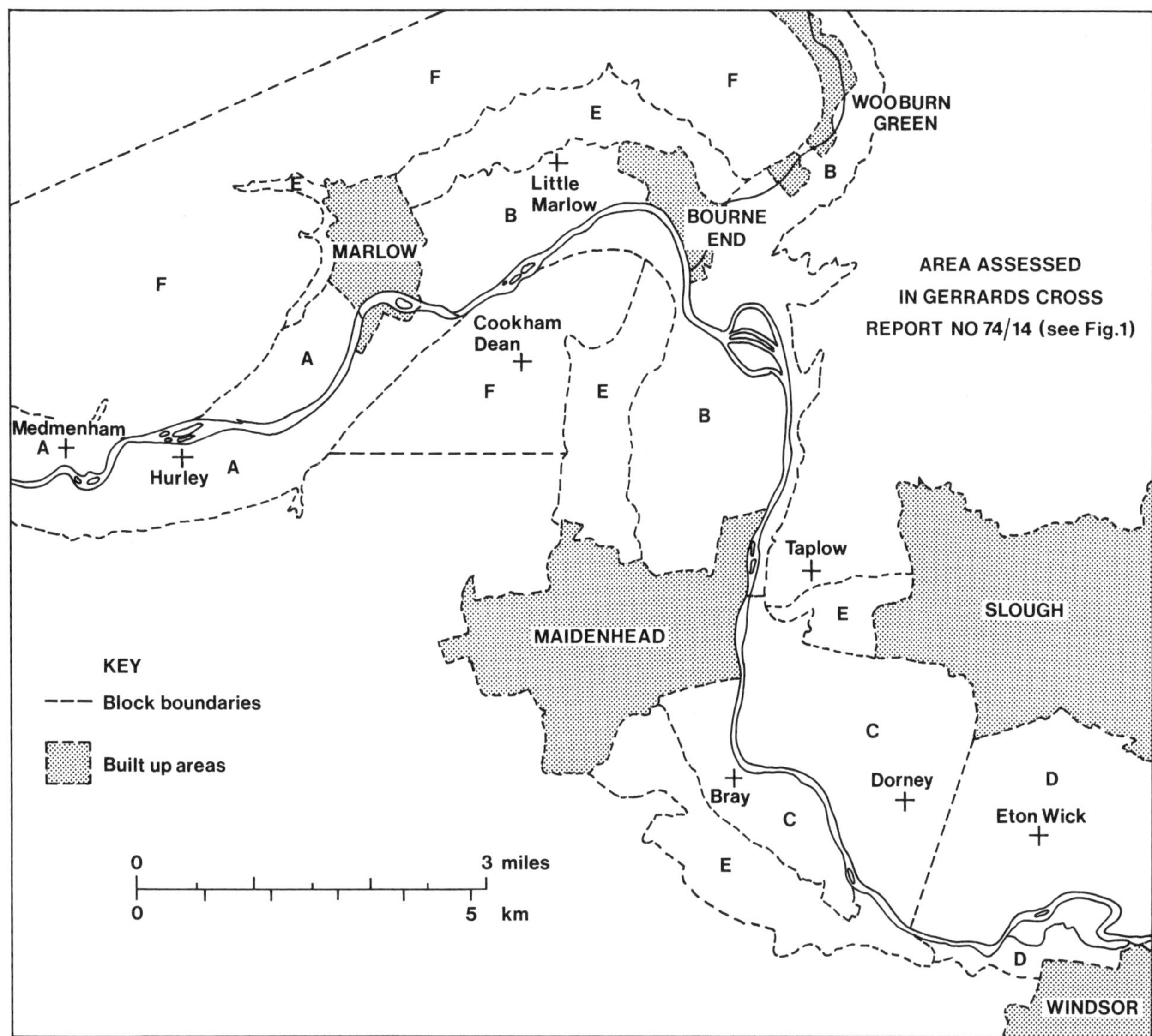

Figure 2 Map showing the location of built-up areas, and the boundaries of the resource blocks

river deposits, and they have only been worked on a small scale locally.

Topography: The topography of the area is dominated by the valley of the River Thames. In the northern half of the area the River Thames and its tributary, the River Wye, have cut deeply into the Chalk producing steep-sided valleys floored by almost flat continuous spreads of alluvium. In the south-east where the river flows over the Reading Beds and London Clay the valley is wider with gentle slopes. The river meanders across the valley floor and has a gentle gradient, falling less than 18 m (40 ft) over a distance of 27 km (17 miles).

The northern part of the area is underlain by chalk and rises to 160 m (525 ft) OD, forming gently sloping downs that are dissected by dry valleys. In the southern part of the area the ground (which is underlain by Reading Beds and London Clay) is lower and gently undulating.

Geology: The area around Maidenhead and Marlow was first surveyed for the Geological Survey on the scale of six inches to one mile by R. L. Sherlock, A. H. Noble, C. N. Bromehead, J. H. Blake and G. Barrow during the period 1902–1920. The Windsor memoir was published in 1915 (Dewey and Bromehead, 1915), and the accompanying one-inch sheet (269) in 1920. Sheet 255 (Beacons-

field) and the accompanying memoir were published in 1922 (Sherlock and Noble, 1922). In 1974 a new edition of the Beaconsfield Sheet was published. During the course of the present survey, the solid and drift lines were amended by H. C. Squirrell. The deposits of the area are classified as shown in Table 1, where they are listed as far as possible in chronological order.

Table 1 Classification of mapped deposits

DRIFT	
Recent and Pleistocene	Alluvium
	Dry Valley Deposits
	River Brickearth
	River Terrace Deposits
	Flood Plain Terrace
	Taplow Terrace
	Boyn Hill Terrace
	Glacial Sand and Gravel
	Clay-with-Flints
	Pebbly Clay and Sand
	Pebble Gravel
SOLID	
Eocene	London Clay
	Reading Beds
Cretaceous	Chalk

3

The solid deposits range in age from the upper part of the Middle Chalk through the Upper Chalk and Reading Beds and into the lower part of the London Clay. The structure is simple; the beds dip gently between 2° and 6° to the south-east and there are no major faults. The Chalk crops out in the north, north-west, south-west and central parts of the area. The Reading Beds form a continuous belt from the north-east to the south-west with some outliers in the central part of the area. The London Clay is mainly confined to the south-east, with some outliers in the central parts of the area.

Chalk: The Middle Chalk outcrops in the deeper valleys to the north of Marlow and Little Marlow. A thickness of up to 25 m crops out in this area and it consists of thickly bedded white chalk containing scattered flint nodules.

The Upper Chalk commences with the Chalk Rock which consists of 2–3 m of thinly bedded hard chalk containing scattered glauconite grains, interbedded with softer nodular chalk. Above the Chalk Rock the sequence continues with massively bedded white chalk containing irregular beds of black flint nodules. The Upper Chalk reaches a maximum thickness of about 90 m in the area.

Reading Beds: The Upper Chalk is unconformably overlain by 15 m of Reading Beds which consist of sands and clays. The beds usually commence with a thin dark clay horizon with flints at the base, passing up into buff-coloured fine to medium sands with thin clay beds. The upper part of the Reading Beds consists of mottled grey, green, red and yellow silty sandy clays with thin beds of clayey sand.

London Clay: The Reading Beds are overlain by the London Clay, which is sandy at the base, passing up into stiff bluish grey silty clays.

The Recent and Pleistocene deposits which contain potentially workable sand and gravel include Alluvium, River Terrace Deposits, Glacial Sand and Gravel, and Pebbly Clay and Sand.

Pebble Gravel: The Pebble Gravel is confined to the highest ground in the north-west, and is composed of pebbles of flint and quartz with minor amounts of quartzite in a sandy matrix. The age and origin of the deposit is uncertain, although White (1906) believed it to be derived from the Reading Beds.

Pebbly Clay and Sand: Pebbly Clay and Sand occurs in small patches in the north of the area. It is composed of well-rounded flint pebbles in a matrix of sand or clay, and according to Sherlock and Noble (1922), it either underlies the Clay-with-Flints or passes laterally into it. Sherlock and Noble considered Pebbly Clay and Sand to represent the more gravelly part of a till derived from the Reading Beds. In Industrial Minerals Assessment Unit borehole 98 NW 66, a thickness of 5.4 m was proved, and this graded as 'clayey' gravel.

Clay-with-Flints: The Clay-with-Flints occurs on the chalk in the hilly areas in the north-west. It consists of brown, reddish brown and yellow silty sandy clays, containing variable amounts of angular-to-rounded pebbles of flint with some quartzite. This deposit is believed to be glacial in origin (Sherlock and Noble, 1922), being the till of an ice sheet that advanced from the west or north-west. In Industrial Minerals Assessment Unit borehole 88 NW 10, a thickness of 3.7 m was proved.

Glacial Sand and Gravel: Glacial Sand and Gravel occurs in the north-west and north-east on the higher ground above the Thames Valley. The deposits consist of poorly sorted sands and gravels that sometimes exhibit bedding and cross-bedding (Sherlock and Noble, 1922). The gravel is composed of subangular to rounded flint pebbles with some quartz, quartzite and sandstone. The proportions of sand, gravel and fines varies considerably, although generally the gravel content exceeds the sand, and the average fines content is 20 per cent. The deposit ranges in recorded thickness from less than 1 m to over 16 m, with an average thickness of some 7 m. The deposits all lie above the highest generally acknowledged terrace of the Thames, and generally to the south-east of the clay-with-flints. Sherlock and Noble (1922) believed these deposits to be of fluvioglacial origin, produced by outwash streams issuing from ice to the north-west and west. In contrast however, Hare (1947) in the light of a detailed geomorphological study believes that the Glacial Sands and Gravels are high-level terrace deposits of the Thames, which originally flowed eastwards from Bourne End through Rickmansworth and Watford. For the sake of expediency the present survey was based upon existing geological maps, which reflect the views of Sherlock and Noble.

Associated with the River Thames and its tributary the Wye, there are river terrace deposits, alluvium, and river brickearths.

Boyn Hill and Taplow terraces: The Boyn Hill and Taplow terraces occur well above the level of the Flood Plain Terrace and can be worked dry. They are covered with poorly sorted sands and gravels of similar composition to the Glacial Sands and Gravel, but with a lower fines content, averaging 13 per cent (Block E and Figure 3). The sands and gravels of these terraces are overlain by thin soils, and occasionally by thin alluvial clays, as found in Industrial Minerals Assessment Unit boreholes 97 NW 64 at Bray and 98 SW 69 at Taplow. In the built-up area of Slough river brickearth rests upon the Taplow Terrace.

Flood Plain Terrace Deposits and Alluvium: These deposits are confined to the immediate vicinity of the river below the level of the Taplow Terrace, and they consist of sandy silty clays (overburden) overlying sand and gravel (mineral). The silts and clays are up to 3.2 m in recorded thickness with an average of 1.4 m. They are pale brownish grey and contain scattered pebbles, occasional shells and local developments of peat. The underlying sand and gravel rests upon bedrock; it is up to 8 m in recorded thickness with a mean of 5.8 m. The gravel is composed predominantly of subangular to rounded flint with well-rounded quartz and quartrite and the sand is composed of flint and quartz. These sands and gravels are similar in composition to the higher terrace gravels; in grading they have a lower fines content with a mean of 5 per cent.

Dry Valley Deposits: In the north-west around Dorney Bottom [953 876] there are dry-valley deposits. These are essentially alluvial deposits, but are called dry-valley deposits because at present the Chalk valley they occupy is dry.

River Brickearth: Resting upon the Flood Plain and Taplow terraces within the built-up area of Slough, upon Boyn Hill Terrace at Lynch Hill Farm [945 825], Burnham, and also upon the Flood Plain Terrace to the east of Marlow, there are spreads of River Brickearth consisting of silty clays.

COMPOSITION OF THE SAND AND GRAVEL DEPOSITS

The potentially workable sand and gravel deposits of the area fall into two categories, namely river deposits and glacial deposits.

River Deposits: The River Deposits consist of alluvium and terrace sands and gravels. The sands and gravels of the Flood Plain Terrace and Alluvium are uniform in composition and grading. The fines content in IMAU borehole samples ranges between 1 and 13 per cent with a mean of 5 per cent, the sand content ranges between 18 and 48 per cent with a mean of 30 per cent, and the gravel content ranges between 45 and 78 per cent with a mean of 65 per cent. The gravel consists of approximately equal proportions of subangular-to-rounded flint with rounded-to-well-rounded quartz and quartzite and minor to trace amounts of limestone, chalk and sandstone. The sands are medium and coarse with very little fine grained material, they are clean, white to pale brown in colour, and are composed of subrounded flint and quartz.

The deposits of the Boyn Hill and Taplow terraces have a higher fines content than those of the Flood Plain Terrace and Alluvium. The fines content ranges between 6 and 19 per cent with a mean of 13 per cent, the sand content ranges between 26 and 47 per cent with a mean of 32 per cent and the gravel content ranges between 41 and 66 per cent with a mean of 55 per cent. The gravel is coarse and fine, composed predominantly of subangular to rounded flint with rounded to well rounded quartz and quartzite and minor to trace amounts of sandstone and chalk. The sand is predominantly medium and coarse grained and is composed of subrounded flint and quartz.

Glacial Deposits: The glacial deposits are more variable in composition than the river deposits. They have a much higher fines content, ranging between 13 and 34 per cent with a mean of 20 per cent. The sand content varies from 20 to 66 per cent with a mean of 42 per cent and the gravel content varies from 14 to 61 per cent with a mean of 43 per cent. The gravel is composed of subangular-to-rounded flint with rounded-to-well-rounded quartz and quartzite and minor amounts of chalk and sandstone; the coarse gravel fraction generally predominates over the fine. The sand is predominantly medium and coarse, being composed of subrounded quartz and flint.

The Flood Plain Terrace deposits and alluvium make up resource blocks A, B, C and D, the Boyn Hill and Taplow terrace deposits make up block E, and the Glacial Sand and Gravel deposits block F. A comparison of the grading characteristics of the various deposits can be made from Figure 3 and Tables 2 to 8. These indicate a general decrease in the fines content from the Glacial Sand and Gravel deposits through the Boyn Hill and Taplow terrace deposits to the Flood Plain Terrace deposits and a corresponding increase in the gravel content. The River Brickearths and Clay-with-Flints do not contain sand and gravel and although the Pebble Gravel may contain mineral, it is of limited extent and therefore has not been assessed.

THE MAP
The sand and gravel resource map is folded into the pocket at the end of this report. The base map is the Ordnance Survey 1:25 000 Outline Edition in grey, on which the topography is shown by contours in green, the geological data in black and the mineral resource information in shades of red.

Geological data: The geological boundary lines are taken from geological maps of this area, which was originally surveyed at the scale of 1:10 560 between 1902 and 1920, and amended in the light of more recent information. Borehole data, which include the stratigraphic relations and mean particle size distribution of the sand and gravel samples collected during the assessment survey, are also shown.

The geological boundaries represent the best available interpretation of the information available at the time of the survey. However, it is inevitable, particularly with drift deposits which change rapidly both vertically and laterally, that local discrepancies may occur.

Mineral resource information: The mineral-bearing ground is subdivided into resource blocks (see Appendix A). Within a resource block the mineral is subdivided into areas where it is 'exposed' and areas where it is present in continuous spreads beneath overburden. The mineral is identified as exposed where the overburden averages less than 1.0 m (3.5 ft) in thickness.

Areas where bedrock outcrops and where boreholes indicate absence of sand and gravel (mineral) are uncoloured on the map. Areas of unassessed sand and gravel, for example, in built-up areas, are indicated by a red stipple.

The area of the exposed sand and gravel is measured from the mapped geological boundary lines. The whole of the area is considered as mineral, although it may include small areas where sand and gravel is not present or is not potentially workable. Inferred boundaries (for which a distinctive symbol is used) have been inserted between categories of deposits. Such boundaries are drawn primarily for the purpose of volume estimation. For the purpose of measuring areas the centre-line of the symbol is used.

RESULTS
The statistical results are summarised in Table 8. Fuller particulars of grading are shown in Figure 3.

For the six resource blocks the confidence limits at the symmetrical 95 per cent probability level (Table 8) vary between 10 per cent and 40 per cent (that is, it is probable that 19 times out of 20 the volumes present lie within these limits). However, the true values are more likely to be nearer the figures estimated than the limits. Moreover, it is probable that in each block roughly the same percentage limits would apply for the estimate of volume of a very much smaller parcel of ground (say, 100 hectares) containing similar sand and gravel deposits if the results from the same number of sample points (as provided by, say, 10 boreholes) were used in the calculation. Thus if closer limits are needed for quotation of reserves of part of a block, it can be expected that data from more than 10 sample points will be required, even if the area is quite small. This point can be illustrated by

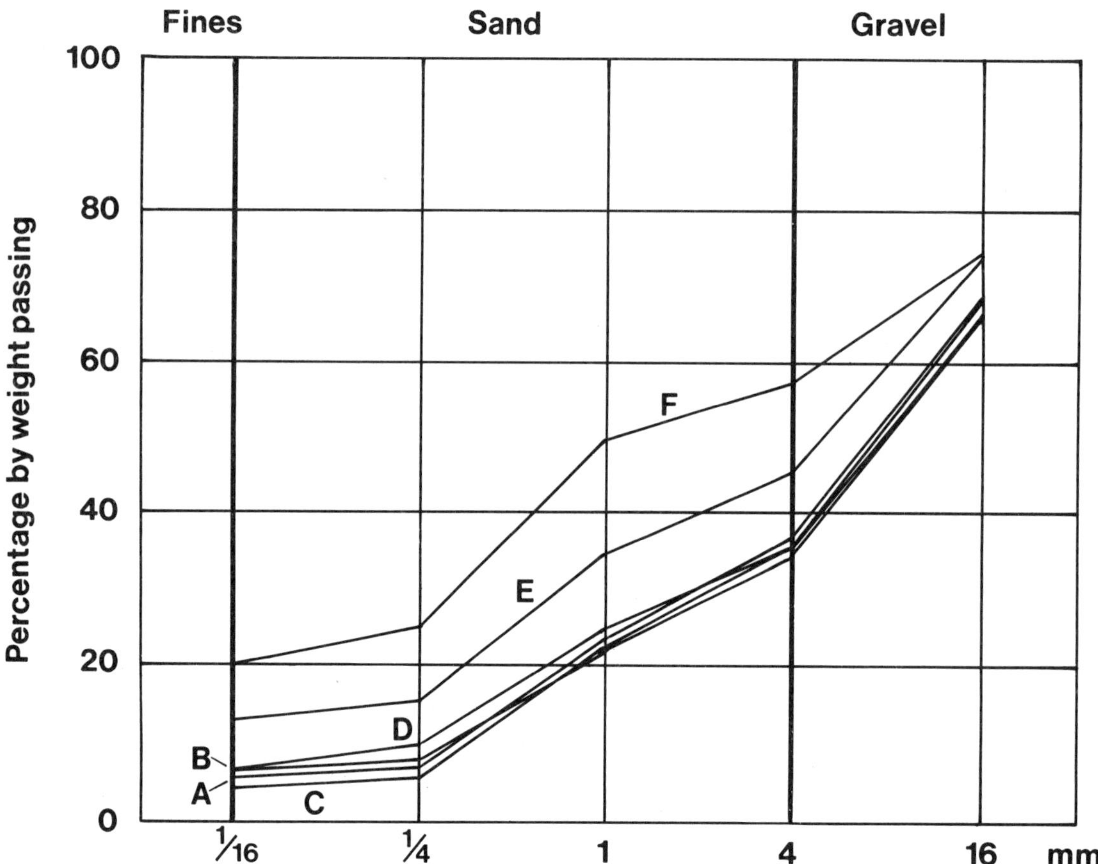

Figure 3 Particle size distribution for the assessed thickness of sand and gravel in resource blocks A to F

Block	Percentage by weight passing				
	$\frac{1}{16}$ mm	$\frac{1}{4}$ mm	1 mm	4 mm	16 mm
A	5	7	23	37	68
B	6	8	22	34	66
C	4	6	23	35	68
D	6	9	25	35	66
E	13	16	35	45	73
F	20	25	50	57	74

considering the whole of the potentially workable sand and gravel on this sheet. The volume (357 million m³) can be estimated to the limits of ±8 per cent at the 95 per cent probability level, by a calculation based upon the data from 78 sample points spread across the six resource blocks.

However, it must be emphasised that the quoted volume of sand and gravel has no simple relationship to the amount that could be extracted in practice, as no allowance has been made in the calculations for any restraints (such as existing buildings and roads) on the use of the land for mineral working.

NOTES ON RESOURCE BLOCKS

The area is divided into six resource blocks (blocks A–F), the boundaries of which are determined as far as possible by geological considerations. The deposits of the Flood Plain Terrace and Alluvium of the Thames have been divided into four blocks designated A, B, C and D. The higher river terraces (Boyn Hill and Taplow) are grouped together in Block E, and the Glacial Sand and Gravel in Block F.

Block A
This block comprises an area of 9.1 km² (all of which is mineral-bearing) situated on the flood plain of the Thames to the west of Marlow [850 865]. The area is flat and low lying, and the deposits have not been worked.

The assessment of resources is based upon 9 Industrial Minerals Assessment Unit boreholes (Table 2) and 11 other records. The overburden ranges in thickness from 0.8 m in borehole 88 SW 23 to 2.2 m in borehole 88 SW 20 (Table 2). It has an average thickness of 1.4 m and usually consists of silty clays with scattered pebbles. The thickness of mineral ranges from 4.9 m in borehole 88 SW 19 to 9.9 m in borehole 88 SW 18, and the mean thickness is 6.2 m. The estimated volume of mineral is 56 million m³ ± 9 million m³ at the 95 per cent confidence level.

The grading results indicate that the deposits in all the boreholes are gravel. The fines content is uniformly low; ranging from 2 per cent in boreholes 88 SW 20 and

Table 2 Data from assessment boreholes: Block A

| Borehole | Recorded thickness | | Mean grading percentage | | | | | |
	Mineral	Over burden	Fines $-\frac{1}{16}$ mm	Fine Sand $-\frac{1}{4}+\frac{1}{16}$ mm	Medium Sand $-1+\frac{1}{4}$ mm	Coarse Sand $-4+1$ mm	Fine gravel $-16+4$ mm	Coarse gravel $+16$ mm
	m	m						
88 SW 18	9.9	1.4	7	2	15	18	37	21
88 SW 19	4.9	1.5	5	2	16	6	36	35
88 SW 20	6.7	2.2	2	1	10	9	34	44
88 SW 21	5.6	1.0	5	3	16	16	32	28
88 SW 22	5.8	1.3	3	2	18	15	30	32
88 SW 23	5.5	0.8	7	1	23	15	28	26
88 NW 19	5.3	1.5	10	2	16	16	24	32
88 NE 15	7.6	1.5	2	1	18	15	25	39
88 NE 18	5.6	1.8	4	1	9	13	36	37

88 NE 15, to 10 per cent in 88 NW 19. The sand content varies from 20 per cent in borehole 88 SW 20 to 39 per cent in borehole 88 SW 23, and it is composed of approximately equal proportions of medium and coarse fractions with a characteristically low proportion of fine sand. The gravel content ranges from 54 per cent in borehole 88 SW 23 to 78 per cent in borehole 88 SW 20. It is composed of equal proportions of coarse and fine material consisting of subangular to rounded flint with rounded to well rounded quartz and quartzite. The mean grading for the block is fines 5 per cent, sand 32 per cent and gravel 63 per cent.

Block B
This block covers an area of 15.7 km² of which 11.2 km² is mineral-bearing. The block is situated along the flood plain of the River Thames between Marlow and Maidenhead, and it also includes the flood plain of the River Wye. There are extensive sand and gravel workings between Marlow and Bourne End [890 875] and there is also a pit [895 827] immediately to the north of Maidenhead.

The assessment of resources is based upon 10 Industrial Minerals Assessment Unit boreholes (Table 3) and 44 other records. The overburden, consisting of silts and clays with scattered pebbles, ranges in thickness from 0.3 m in borehole 98 NW 71 to 3.2 m in borehole 98 NW 69 (Table 3), and it has a mean thickness of 1.2 m. In the Thames valley the overburden is usually over a metre thick, while in the Wye valley the thickness is considerably less than a metre. An inferred boundary (see map) has been inserted between these two categories of deposits. The thickness of mineral varies from 3.9 m in borehole 88 SE 27 to 8.0 m in borehole 88 NE 23, and the mean thickness is 5.7 m. The estimated volume of mineral within the block is 64 million m³ ± 6 million m³ at the 95 per cent confidence level.

The grading results indicate 'clayey' gravel in boreholes 88 NE 28 and 98 NW 69, and gravel in all the other boreholes. The fines content varies between 2 per cent in boreholes 88 NE 23 and 88 NE 20 to 13 per cent in borehole 98 NW 69. The sand content varies from 21 per cent in boreholes 88 SE 27 and 98 NW 70, to 34 per cent in boreholes 88 NE 23 and 88 NE 28, and it is composed of approximately equal proportions of the coarse and medium fractions with only minor amounts of fine sand. The gravel content ranges from 55 per cent in borehole 88 NE 28 to 76 per cent in borehole 88 SE 27, and is composed of approximately equal proportions of coarse and fine material consisting predominantly of subangular to rounded flint with rounded quartz and quartzite. The mean grading for the block is fines 6 per cent, sand 28 per cent and gravel 66 per cent.

Table 3 Data from assessment boreholes: Block B

| Borehole | Recorded thickness | | Mean grading percentage | | | | | |
	Mineral	Over-burden	Fines $-\frac{1}{16}$ mm	Fine sand $-\frac{1}{4}+\frac{1}{16}$ mm	Medium sand $-1+\frac{1}{4}$ mm	Coarse sand $-4+1$ mm	Fine gravel $-16+4$ mm	Coarse gravel $+16$ mm
	m	m						
88 NE 20	6.6	1.6	2	1	13	10	29	45
88 NE 21	5.6	1.3	9	2	15	9	30	35
88 NE 23	8.0	0.5	2	2	19	13	29	35
88 NE 28	7.2	0.7	11	1	17	16	30	25
88 SE 27	3.9	2.2	3	1	10	10	41	35
98 NW 68	5.5	0.6	5	1	10	11	31	42
98 NW 69	5.6	3.2	13	2	15	14	28	28
98 NW 70	5.7	1.3	4	2	8	11	45	30
98 NW 71	4.7	0.3	4	1	13	14	28	40
98 SW 67	5.6	0.8	5	3	16	11	33	32

Table 4 Data from assessment boreholes: Block C

| Borehole | Recorded thickness | | Mean grading percentage | | | | | |
	Mineral	Over-burden	Fines $-\frac{1}{16}$ mm	Fine sand $-\frac{1}{4}+\frac{1}{16}$ mm	Medium sand $-1+\frac{1}{4}$ mm	Coarse sand $-4+1$ mm	Fine gravel $-16+4$ mm	Coarse gravel $+16$ mm
	m	m						
87 NE 34	5.2	1.5	7	6	36	6	20	25
97 NW 60	7.8	0.6	7	3	21	13	29	27
97 NW 61	6.1	1.8	1	0	9	12	41	37
97 NW 63	5.3	2.4	1	1	12	14	38	34
97 NW 65	4.8	0.9	6	2	14	12	30	36
97 NW 66	5.6	1.3	3	1	12	12	37	35
98 SW 68	4.4	2.5	3	1	10	10	33	43
98 SW 70	4.2	0.8	7	3	23	11	36	20

Block C

This block is situated to the south-east of Maidenhead, along the flood plain of the River Thames. The block covers an area of 11.7 km² of which 10.8 km² is mineral bearing. There are two gravel pits within the block, one south of Taplow [910 810], and the other south-east of Bray [913 783].

The assessment of resources is based upon 8 Industrial Minerals Assessment Unit boreholes (Table 4) and 40 other records. The overburden, consisting of silts and clays with scattered pebbles, ranges in thickness from 0.6 m in borehole 97 NW 60 to 2.5 m in borehole 98 SW 68 (Table 4), and it has a mean thickness of 1.4 m. The mineral ranges in thickness from 4.2 m in borehole 98 SW 70 to 7.8 m in borehole 97 NW 60 and has a mean of 5.7 m. The estimated volume of mineral is 62 million m³ ± 11 million m³ at the 95 per cent confidence level.

The grading results indicate sandy gravel in borehole 87 NE 34 and gravel in all other boreholes. The fines content ranges from 1 per cent in boreholes 97 NW 61 and 63, to 7 per cent in boreholes 87 NE 34, 97 NW 60 and 98 SW 70. The sand content varies between 21 per cent in boreholes 97 NW 61 and 98 SW 68, and 48 per cent in borehole 87 NE 34; it is composed of medium and coarse sand, with only a small proportion of fine material. The gravel content ranges from 45 per cent in borehole 87 NE 34 to 78 per cent in borehole 97 NW 61, and it consists of equal proportions of fine and coarse material composed of subangular to rounded flint with rounded to well rounded quartz and quartzite. The mean grading for the block is fines 4 per cent, sand 31 per cent and gravel 65 per cent.

Block D

Block D is situated upon the flood plain of the Thames around Eton Wick [950 784] and it covers an area of 9.2 km² of which 9.1 km² is mineral-bearing.

The assessment of resources is based upon 8 Industrial Minerals Assessment Unit boreholes (Table 5) and 2 other records. The overburden, consisting of silts and clays with scattered pebbles, ranges in thickness from 1.0 m in borehole 97 NW 67 to 2.7 m in borehole 97 NE 189 (Table 5) and has a mean of 1.5 m. The thickness of mineral varies from 3.2 m in borehole 97 NE 189 to 6.9 m in borehole 97 NW 67, and it has an average thickness of 5.5 m. The estimated volume of mineral is 51 million m³ ± 10 million m³ at the 95 per cent confidence level.

The grading results indicate 'clayey' gravel in borehole 97 NE 188, and gravel in all the other boreholes. The fines content ranges from 2 per cent in borehole 97 NE 189 to 12 per cent in borehole 97 NE 188. The sand content ranges from 18 per cent in borehole 97 NE 188 to 35 per cent in borehole 27 NW 69, the proportion of fine sand is low, and the medium sand content predominates over the coarse fraction. The gravel content ranges from 58 per cent in borehole 97 NW 69 to 74 per cent in borehole 97 NE 189. The gravel is composed of subangular to rounded flint with rounded quartz and quartzite, and the proportion of the coarse gravel fraction slightly exceeds the fine fraction. The mean grading for the block is, fines 6 per cent, sand 29 per cent and gravel 65 per cent.

Table 5 Data assessment from boreholes: Block D

| Borehole | Recorded thickness | | Mean grading percentage | | | | | |
	Mineral	Over-burden	Fines $-\frac{1}{16}$ mm	Fine sand $-\frac{1}{4}+\frac{1}{16}$ mm	Medium sand $-1+\frac{1}{4}$ mm	Coarse sand $-4+1$ mm	Fine gravel $-16+4$ mm	Coarse gravel $+16$ mm
	m	m						
97 NE 187	6.5	1.3	4	5	19	10	26	36
97 NE 188	4.8	1.9	12	1	9	8	25	45
97 NE 189	3.2	2.7	2	0	14	10	30	44
97 NW 67	6.9	1.0	8	2	18	11	30	31
97 NW 68	5.9	1.6	3	3	18	10	35	31
97 NW 69	5.1	1.4	7	4	22	9	28	30
97 NW 70	4.1	1.7	3	4	15	11	34	33
98 SW 71	5.1	1.9	4	2	13	14	37	30

Table 6 Data from assessment boreholes: Block E

Borehole	Recorded thickness		Mean grading percentage					
	Mineral	Over-burden	Fines $-\frac{1}{16}$ mm	Fine sand $-\frac{1}{4}+\frac{1}{16}$ mm	Medium sand $-1+\frac{1}{4}$ mm	Coarse sand $-4+1$ mm	Fine gravel $-16+4$ mm	Coarse gravel $+16$ mm
	m	m						
87 NE 33	5.2	0.2	8	3	21	10	29	29
88 NE 17	7.2	0.3	15	2	20	10	30	23
88 NE 22	6.8	0.4	16	3	16	9	27	29
88 NE 25	3.7	0.2	19	4	15	7	21	34
88 SE 26	2.1	0.8	6	2	13	13	33	33
97 NW 62	2.9	0.9	9	3	17	13	30	28
97 NW 64	5.1	1.1	12	2	14	10	30	32
98 SW 69	5.5	1.0	12	5	34	8	26	15

Block E

Block E occupies an area of 13.4 km² of which 11.7 km² is mineral-bearing. The block is composed of the Taplow and Boyn Hill terraces of the River Thames; it is divided into five separate areas within the Thames valley above the level of the flood plain. Although deposits of the two terraces are of similar thickness and composition, it is only the Taplow Terrace that has been worked to date. There is one small pit [884 828] immediately to the north of Maidenhead but the main workings are on the Taplow Terrace south-east of Taplow [912 824] and south-west of Bray [902 796].

The assessment of resources is based upon 8 Industrial Minerals Assessment Unit boreholes (Table 6) and 25 other records. The overburden is thin and consists of soil, occasionally underlain by sandy clay; it ranges in thickness from 0.2 m in borehole 88 NE 25 and 87 NE 33 to 1.1 m in borehole 97 NW 64 (Table 6) and has a mean thickness of 0.8 m. The mineral ranges in thickness from 2.1 m in borehole 88 SE 26 to 7.2 m in borehole 88 NE 17 and has a mean thickness of 4.6 m. The estimated volume of mineral is 54 million m³ ± 12 million m³ at the 95 per cent confidence level.

The grading results indicate 'clayey' gravel in borehole 88 NE 17, 22, 25 and 97 NW 64, 'clayey' sandy gravel in borehole 98 SW 69, and gravel in the other three boreholes. The fines content varies between 6 per cent in borehole 88 SE 26 and 19 per cent in borehole 88 NE 25. The sand content ranges from 26 per cent in boreholes 88 NE 25 and 97 NW 64, to 47 per cent in borehole 98 SW 69. The medium sand fraction predominates over the coarse, and fine sand only occurs in minor amounts. The gravel content varies between 41 per cent in borehole 98 SW 69 and 66 per cent in borehole 88 SE 26; it is composed of approximately equal proportions of course and fine material consisting of subangular-to-rounded flint with minor amounts of well-rounded-to-rounded quartz and quartzite. The mean grading for the resource block is fines 13 per cent, sand 32 per cent and gravel 55 per cent.

Block F

This block extends over an area of 33.1 km² of which 10.6 km² is mineral-bearing. It includes the high-level Glacial Sand and Gravel in the north-west of the sheet, extending from Medmenham [805 845] eastwards to Wooburn [910 896], together with occurrences on the south side of the Thames around Cookham Dean. The deposits rest upon the Chalk, usually on the higher ground between the dry valleys at an elevation approximately between 60 m (200 ft) and 120 m (400 ft) above OD.

The assessment of the resources is based upon 12 Industrial Minerals Assessment Unit boreholes (Table 7). The overburden is thin, consisting of soil occasionally underlain by silty sandy clays. The thickness of overburden varies from 0.1 m in boreholes 88 NW 15, 88 NE 19 and 98 NW 66, to 2.2 m in borehole 88 NE 16 (Table 7) and it has a mean thickness of 0.5 m. The

Table 7 Data from assessment boreholes: Block F

Borehole	Recorded thickness		Mean grading percentage					
	Mineral	Over-burden	Fines $-\frac{1}{16}$ mm	Fine sand $-\frac{1}{4}+\frac{1}{16}$ mm	Medium sand $-1+\frac{1}{4}$ mm	Coarse sand $-4+1$ mm	Fine gravel $-16+4$ mm	Coarse gravel $+16$ mm
	m	m						
88 NW 11	7.0	0.2	21	1	15	9	20	34
88 NW 13	5.0	0.3	20	13	49	4	5	9
88 NW 15	6.8	0.1	20	5	36	8	14	17
88 NW 16	1.6	1.9	34	9	30	4	5	18
88 NW 17	3.8	0.2	17	3	15	7	23	35
88 NW 18	5.2	0.3	18	3	26	10	23	20
88 NE 14	5.4	0.3	19	3	10	7	19	42
88 NE 16	11.5	2.2	26	2	11	7	19	35
88 NE 19	4.5	0.1	13	3	16	8	22	38
88 NE 24	5.9	0.3	14	4	18	10	29	25
88 NE 26	16.3	0.4	21	8	30	5	13	23
98 NW 66	8.4	0.1	15	7	41	5	13	19

mineral is of variable thickness, ranging from 1.6 m in
borehole 88 NW 16 to 16.3 m in borehole 88 NE 26 with
a mean of 6.5 m. The estimated volume of mineral is
70 million m³ ± 28 million m³ at the 95 per cent confidence level.

The grading results indicate that the deposits are very
variable; all have a high fines content, ranging from
'clayey' gravel through 'very clayey' gravel and 'very
clayey' sandy gravel to 'very clayey' pebbly sand. The
fines content varies between 13 per cent in borehole
88 NE 19 and 34 per cent in borehole 88 NW 16. The
sand content ranges from 20 per cent in boreholes
88 NE 14 and 16 to 66 per cent in borehole 88 NW 13.
The sand is mainly medium grained with small amounts
of fine and coarse. The gravel content varies from 14 per
cent in borehole 88 NW 13 to 61 per cent in borehole
88 NE 14. The gravel is coarse with fine, and is composed of angular to rounded flint with rounded to well
rounded quartz and quartzite and minor to trace amounts
of chalk and sandstone. The mean grading for the block
is fines 20, sand 37 and gravel 43 per cent.

Table 8 Summary of statistical results

Block	Area		Mean thickness				Volume of mineral				Mean grading percentage		
	Block	Mineral	Over-burden		Mineral				Limits at the 95% confidence level		Fines	Sand	Gravel
	km²	km²	m	ft	m	ft	m³× 10⁶	yd³× 10⁶	±%	±m³× 10⁶	$-\frac{1}{16}$ mm	$-4+\frac{1}{16}$	+4 mm
A	9.1	9.1	1.4	4.5	6.2	20.5	56	74	16	9	5	32	63
B	15.7	11.2	1.2	4.0	5.7	18.5	64	84	10	6	6	28	66
C	11.7	10.8	1.4	4.5	5.7	18.5	62	80	17	11	4	31	65
D	9.2	9.1	1.5	5.0	5.5	18.5	51	66	19	10	6	29	65
E	13.4	11.7	0.8	2.5	4.6	15.0	54	71	23	12	13	32	55
F	33.1	10.6	0.5	1.5	6.5	21.5	70	91	40	28	20	37	43
A to F	92.2	62.5	1.1	3.5	5.7	18.5	357	466	8	30	9	31	60

APPENDIX A

FIELD AND LABORATORY PROCEDURES

Trial and error during initial studies of the complex and variable glacial deposits of East Anglia and Essex showed that an absolute minimum of five sample points evenly distributed across the sand and gravel are needed to provide a worthwhile statistical assessment, but that, where possible, there should be not less than ten. Sample points are any points for which adequate information exists about the nature and thickness of the deposit and may include boreholes other than those drilled during the survey and exposures. In particular, the cooperation of sand and gravel operators ensures that boreholes are not drilled where reliable information is already available; although this may be used in the calculations, it is held confidentially by the Institute and cannot be disclosed.

The mineral shown on each 1:25 000 sheet is divided into resource blocks. The arbitrary size selected, 10 km², is a compromise to meet the aims of the survey by providing sufficient sample points in each block. As far as possible the block boundaries are determined by geological boundaries so that, for example, glacial and river terrace gravels are separated. Otherwise division is by arbitrary lines, which may bear no relationship to the geology. The blocks are drawn provisionally before drilling begins.

A reconnaissance of the ground is carried out to record any exposures and inquiries are made to ascertain what borehole information is available. Borehole sites are then selected to provide an even pattern of sample points at a density of approximately one per square kilometre. However, because broad trends are independently overlain by smaller scale characteristically random variations, it is unnecessary to adhere to a square grid pattern. Thus such factors as ease of access and the need to minimise disturbance to land and the public are taken into account in siting the holes; at the same time it is necessary to guard against the possibility that ease of access (that is, the positions of roads and farms) may reflect particular geological conditions, which may bias the drilling results.

The drilling machine employed should be capable of providing a continuous sample representative of all unconsolidated deposits, so that the in-situ grading can be determined, if necessary, to a depth of 30 m (100 ft) at a diameter of about 200 mm (8 in), beneath different types of overburden. It should be reliable, quiet, mobile and relatively small (so that it can be moved to sites of difficult access). Shell and auger rigs have proved to be almost ideal.

The rigs are modified to enable deposits above the water table to be drilled 'dry', instead of with water added to facilitate the drilling, to minimise the amount of material drawn in from outside the limits of the hole. The samples thus obtained are representative of the in-situ grading, and satisfy one of the most important aims of the survey. Below the water table the rigs are used conventionally, although this may result in the loss of some of the fines fraction and the pumping action of the bailer tends to draw unwanted material into the hole from the sides or the bottom.

A continuous series of bulk samples is taken throughout the sand and gravel. Ideally samples are composed exclusively of the whole of the material encountered in the borehole between stated depths. However, care is taken to discard, as far as possible, material which has caved or has been pumped from the bottom of the hole. A new sample is commenced whenever there is an appreciable lithological change within the sand and gravel, or at every 1 m (3.3 ft) depth. The samples, each weighing between 25 and 45 kg (55 and 100 lb), are despatched in heavy duty polythene bags to a laboratory for grading. The grading procedure is based on British Standard 1377 (1967). Random checks on the accuracy of the grading are made in the Institute's laboratories.

All data, including mean grading analysis figures calculated for the total thickness of the mineral, are entered on standard record sheets, abbreviated copies of which are reproduced in Appendix F.

Detailed records may be consulted at the appropriate offices of the Institute, upon application to the Head, Industrial Minerals Assessment Unit.

APPENDIX B

STATISTICAL PROCEDURE

Statistical assessment

1 A statistical assessment is made of an area of mineral greater than 2 km², if there is a minimum of five evenly spaced boreholes in the resource block (for smaller areas see paragraph 12 below).

2 The simple methods used in the calculations are consistent with the amount of data provided by the survey. Conventional symmetrical confidence limits are calculated for the 95 per cent probability level, that is, there is a 5 per cent or one in twenty chance of a result falling outside the stated limits.

3 The volume estimate (V) for the mineral in a given block is the product of the two variables, the sampled areas (A) and the mean thickness ($\bar{l}_m$) calculated from the individual thicknesses at the sample points. The standard deviations for these variables are related such that

$$S_V = \sqrt{(S_A{}^2 + S_{\bar{l}m}{}^2)} \quad . \tag{1}$$

4 The above relationship may be transposed such that

$$S_V = S_{\bar{l}m} \sqrt{(1 + S_A{}^2/S_{\bar{l}m}{}^2)} \quad . \tag{2}$$

From this it can be seen that as $S_A{}^2/S_{\bar{l}m}{}^2$ tends to 0, S_V tends to $S_{\bar{l}m}$.

If, therefore, the standard deviation for area is small with respect to that for mean thickness, the standard deviation for volume approximates to that for mean thickness.

5 Given that the number of approximately evenly spaced sample points in the sampled area is n with mineral thickness measurements $l_{m_1}, l_{m_2}, \ldots l_{m_n}$, then the best estimate of mean thickness, $\bar{l}_m$, is given by

$$\Sigma(l_{m_1} + l_{m_2} \ldots l_{m_n})/n.$$

For groups of closely spaced boreholes a discretionary weighting factor may be applied to avoid bias (see note on weighting below). The standard deviation for mean thickness $S_{\bar{l}}$, expressed as a proportion of the mean thickness is given by

$$S_{\bar{l}} = (1/\bar{l}_m) \sqrt{[l_m - \bar{l}_m)^2/(n - 1)]}$$

where l_m is any value in the series l_{m_1} to l_{m_n}.

6 The sampled area in each resource block is coloured pink on the map. Wherever possible, calculations relate to the mineral within mapped geological boundaries (which may not necessarily correspond to the limits of deposit). Where the area is not defined by a mapped boundary, that is, where the boundary is inferred, a distinctive symbol is used. Experience suggests that the errors in determining area are small relative to those in thickness. The relationship $S_A/S_{\bar{l}m} \leqslant \frac{1}{3}$ is assumed in all cases. It follows from equation [2] that

$$S_{\bar{l}m} \leqslant S_V \leqslant 1.05 \, S_{\bar{l}m} \quad . \tag{3}$$

7 The limits on the estimate of mean thickness of mineral, $L_{\bar{l}m}$, may be expressed in absolute units $\pm (t/\sqrt{n}) \times S_{\bar{l}m}$ or as a percentage $\pm (t/\sqrt{n}) \times S_{\bar{l}m} \times (100/\bar{l}_m)$ per cent, where t is Student's t at the 95 per cent probability level for $(n - 1)$ degrees of freedom, evaluated by reference to statistical tables. (In applying Student's t it is assumed that the measurements are distributed normally).

Block calculation **1:25 000** $\left.\begin{array}{l}\\ \text{Block}\end{array}\right\}$ Fictitious

Area
Block: 11.08 km²
Mineral: 8.32 km²

Mean thickness
Overburden: 2.5 m
Mineral: 6.5 m

Volume
Overburden: 21 million m³
Mineral: 54 million m³
Confidence limits of the estimate of mineral volume at the
95 per cent probability level: ± 20 per cent
That is, the volume of mineral (with 95 per cent probability):
54 ± 11 million m³
Thickness estimate measurements in metres
l_o = overburden thickness l_m = mineral thickness

Sample point	Weighting w	Overburden		Mineral		Remarks
		l_o	wl_o	l_m	wl_m	
SE 14	1	1.5	1.5	9.4	9.4	
SE 18	1	3.3	3.3	5.8	5.8	
SE 20	1	nil	–	6.9	6.9	IMAU boreholes
SE 22	1	0.7	0.7	6.4	6.4	
SE 23	1	6.2	6.2	4.1	4.1	
SE 24	1	4.3	4.3	6.4	6.4	
SE 17	½	1.2 } 1.6		9.8 } 7.2		Hydrogeology Unit record
123/45	½	2.0		4.6		
1	¼	2.7 } 2.6		7.3 } 5.8		Close group of four boreholes (commercial)
2	¼	4.5		3.2		
3	¼	0.4		6.8		
4	¼	2.8		5.9		
Totals	$\Sigma w = 8$	$\Sigma wl_o = 20.2$		$\Sigma wl_m = 52.0$		
Means	—	$\bar{l}_o = 2.5$		$\bar{l}_m = 6.5$		

Calculation of confidence limits

l_m	$(l_m - \bar{l}_m)$	$(l_m - \bar{l}_m)^2$
9.4	2.9	8.41
5.8	0.7	0.49
6.9	0.4	0.16
6.4	0.1	0.01
4.1	2.4	5.76
6.4	0.1	0.01
7.2	0.7	0.49
5.8	0.7	0.49

$\Sigma(l_m - \bar{l}_m)^2 = 15.82$
$n = 8$
$t = 2.365$

L_V is calculated as

$1.05\,(t/\bar{l}_m)\,\sqrt{[\,\Sigma(l_m - \bar{l}_m)^2/n(n - 1)]} \times 100$
$= 1.05 \times (2.365/6.5)\,\sqrt{[15.82/(8 \times 7)]} \times 100$
$= 20.3$
$\simeq 20\,\text{per cent}.$

Figure 4 Example of resource block assessment: calculation and results

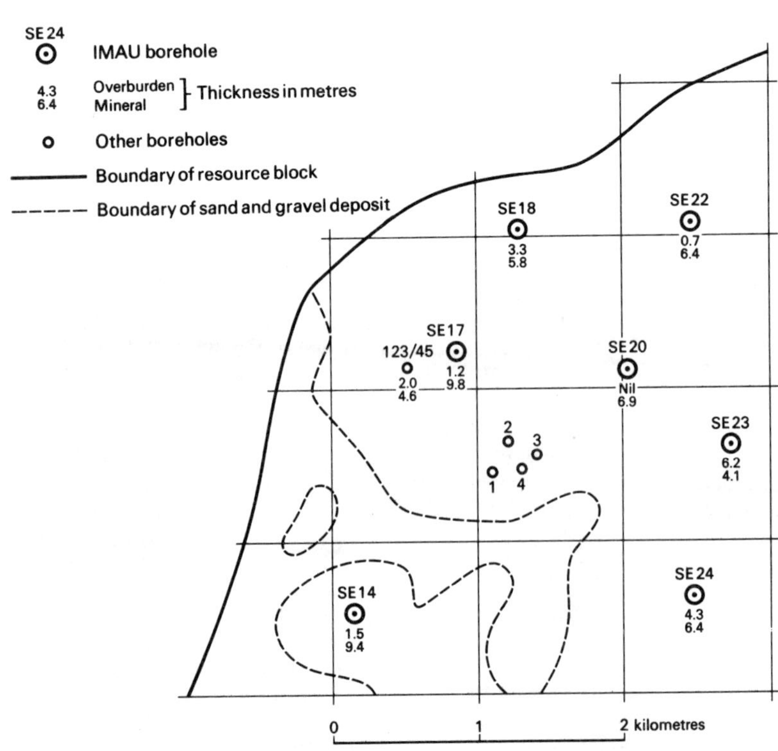

Figure 5 Example of resource block assessment: map of a fictitious block

8   Values of t at the 95 per cent probability level for values of n up to 20 are as follows:

n	t	n	t
1	infinity	11	2.228
2	12.706	12	2.201
3	4.303	13	2.179
4	3.182	14	2.160
5	2.776	15	2.145
6	2.571	16	2.131
7	2.447	17	2.120
8	2.365	18	2.110
9	2.306	19	2.101
10	2.262	20	2.093

(from Table 12, Biometrika Tables for Statisticians, Volume 1, Second Edition, Cambridge University Press, 1962). When n is greater than 20, 1.96 is used (the value of t when n is infinity).

9   In calculating confidence limits for volume, L_V, the following inequality corresponding to equation [3] is applied: $L_{l\mathrm{m}} \leqslant L_V \leqslant 1.05\, L_{\bar{l}\mathrm{m}}$

10   In summary, for values of n between 5 and 20, L_V is calculated as

$$[(1.05 \times t)/\bar{l}_\mathrm{m}] \times [\sqrt{} \ \Sigma(l_\mathrm{m} - \bar{l}_\mathrm{m})^2/n(n-1)] \times 100$$

per cent, and when n is greater than 20, as

$$[(1.05 \times 1.96)/\bar{l}_\mathrm{m}] \times [\sqrt{} \ \Sigma(l_\mathrm{m} - \bar{l}_\mathrm{m})^2/n(n-1)] \times 100$$

per cent.

11   The application of this procedure to a fictitious area is illustrated in Figures 4 and 5

Inferred assessment
12   If the sampled area of mineral in a resource block is between 0.25 km² and 2 km² an assessment is inferred, based on geological and topographical information usually supported by the data from one or two boreholes. The volume of mineral is calculated as the product of the area, measured from field data, and the estimated thickness. Confidence limits are not calculated.

13   In some cases a resource block may include an area left uncoloured on the map, within which mineral (as defined) is interpreted to be generally absent. If there is reason to believe that some mineral may be present, an inferred assessment may be made.

14   No assessment is attempted for an isolated area of mineral less than 0.25 km².

15   *Note on weighting*   The thickness of a deposit at any point may be governed solely by the position of the point in relation to a broad trend. However, most sand and gravel deposits also exhibit a random pattern of local, and sometimes considerable, variation in thickness. Thus the distribution of sample points need be only approximately regular and in estimating the mean thickness only simple weighting is necessary. In practice, equal weighting can often be applied to thicknesses at all sample points. If, however, there is a distinctly unequal distribution of points, bias is avoided by dividing the sampled area into broad zones, to each of which a value roughly proportional to its area is assigned. This value is then shared between the data points within the zone as the weighting factor.

CLASSIFICATION AND DESCRIPTION OF SAND AND GRAVEL

For the purposes of assessing resources of sand and gravel a classification should take account of economically important characteristics of the deposit, in particular the absolute content of fines and the ratio of sand to gravel.

The terminology commonly used by geologists when describing sedimentary rocks (Wentworth, 1922) is not entirely satisfactory for this purpose. For example, Wentworth proposed that a deposit should be described as a 'gravelly sand' when it contains more sand that gravel and there is at least 10 per cent of gravel, provided that there is less than 10 per cent of material finer than sand (less than $\frac{1}{16}$ mm) and coarser than pebbles (more than 64 mm in diameter). Because deposits containing more than 10 per cent fines are not embraced by this system a modified binary classification based on Willman (1942) has been adopted.

When the fines content exceeds 40 per cent the material is not considered to be potentially workable and falls outside the definition of mineral. Deposits which contain 40 per cent fines or less are classified primarily on the ratio of sand to gravel but qualified in the light of the fines content, as follows: less than 10 per cent fines—no qualification; 10 per cent or more but less than 20 per cent fines—'clayey'; 20 to 40 per cent fines—'very clayey'.

The term 'clay' (as written, with single quote marks) is used to describe all material passing $\frac{1}{16}$ mm. Thus it has no mineralogical significance and includes particles falling within the size range of silt. The normal meaning applies to the term clay where it does not appear in single quotation marks.

The ratio of sand to gravel defines the boundaries between sand, pebbly sand, sandy gravel and gravel (at 19:1, 3:1 and 1:1).

Thus it is possible to classify the mineral into one of twelve descriptive categories (see Figure 6). The procedure is as follows:
1   Classify according to ratio of sand to gravel.
2   Describe fines.

For example, a deposit grading 11 per cent gravel, 70 per cent sand and 19 per cent fines is classified as 'clayey' pebbly sand. This short description is included in the borehole log (see Note 11, p. 16).

Many differing proposals exist for the classification of the grain size of sediments (Atterberg, 1905; Udden, 1914; Wentworth, 1922; Wentworth, 1935; Allen, 1936; Twenhofel, 1937; Lane and others, 1947). As Archer (1970a, b) has emphasised, there is a pressing need for a simple metric scale acceptable to both scientific and engineering interests, for which the class limit sizes correspond closely with certain marked changes in the natural properties of mineral particles. For example, there is an important change in the degree of cohesion between particles at about the $\frac{1}{16}$-mm size, which approximates to the generally accepted boundary between silt and sand. These and other requirements are met by a system based on Udden's geometric scale and a simplified form of Wentworth's terminology (Table 9), which is used in this Report.

The fairly wide intervals in the scale are consistent with the general level of accuracy of the qualitative assessments of the resource blocks. Three sizes of sand are recognised, fine ($-\frac{1}{4} + \frac{1}{16}$ mm), medium ($-1 + \frac{1}{4}$ mm) and coarse ($-4 + 1$ mm). The boundary at 16 mm distinguishes a range of finer gravel ($-16 + 4$ mm), often characterised by abundance of worn tough pebbles of vein quartz, from larger pebbles often of notably different materials. The boundary at 64 mm distinguishes pebbles from cobbles. The term 'gravel' is used loosely to denote both pebble-sized and cobble-sized material.

The size distribution of borehole samples is determined by sieve analysis, which is presented by the laboratory as logarithmic cumulative curves (see, for example, British

Standard 1377: 1967). In this report the grading is tabulated on the borehole record sheets (Appendix F), the intercepts corresponding with the simple geometric scale $\frac{1}{16}$ mm, $\frac{1}{4}$ mm, 1 mm, 4 mm, 16 mm and so on as required. Original sample grading curves are available for reference at the appropriate office of the Institute.

Each bulk sample is described, subjectively, by a geologist at the borehole site. Being based on visual examination, the description of the grading is inexact, the accuracy depending on the experience of the observer. The descriptions recorded are modified, as necessary, when the laboratory results become available.

The relative proportions of the rock types present in the gravel fraction are indicated by the use of the words 'and' or 'with'. For example, 'flint and quartz' indicates very approximate equal proportions with neither constituent accounting for less than about 25 per cent of the whole; 'flint with quartz' indicates that flint is dominant and quartz, the principal accessory rock type, comprises 5 to 25 per cent of the whole. Where the accessory material accounts for less than 5 per cent of the whole, but is still readily apparent, the phrase 'with some' has been used. Rare constituents are referred to as 'trace'.

The terms used in the field to describe the degree of rounding of particles, which is concerned with the sharpness of the edges and corners of a clastic fragment and not the shape (after Pettijohn, 1957), are as follows.

Angular: showing little or no evidence of wear; sharp edges and corners.

Subangular: showing definite effects of wear. Fragments still have their original form but edges and corners begin to be rounded off.

Subrounded: showing considerable wear. The edges and corners are rounded off to smooth curves. Original grain shape is still distinct.

Rounded: original faces almost completely destroyed, but some comparatively flat surfaces may still remain. All original edges and corners have been smoothed off to rather broad curves. Original shape is still apparent.

Well-rounded: no original faces, edges or corners left. The entire surface consists of broad curves; flat areas are absent. The original shape is suggested by the present form of the grain.

Table 9 Classification of gravel, sand and fines

Size limits	Grain size description	Qualification	Primary classification
	Cobble		
64 mm –			
		Coarse	Gravel
16 mm –	Pebble		
		Fine	
4 mm –			
		Coarse	
1 mm –			
	Sand	Medium	Sand
$\frac{1}{4}$ mm –			
		Fine	
$\frac{1}{16}$ mm –			
	Fines (silt and clay)		Fines

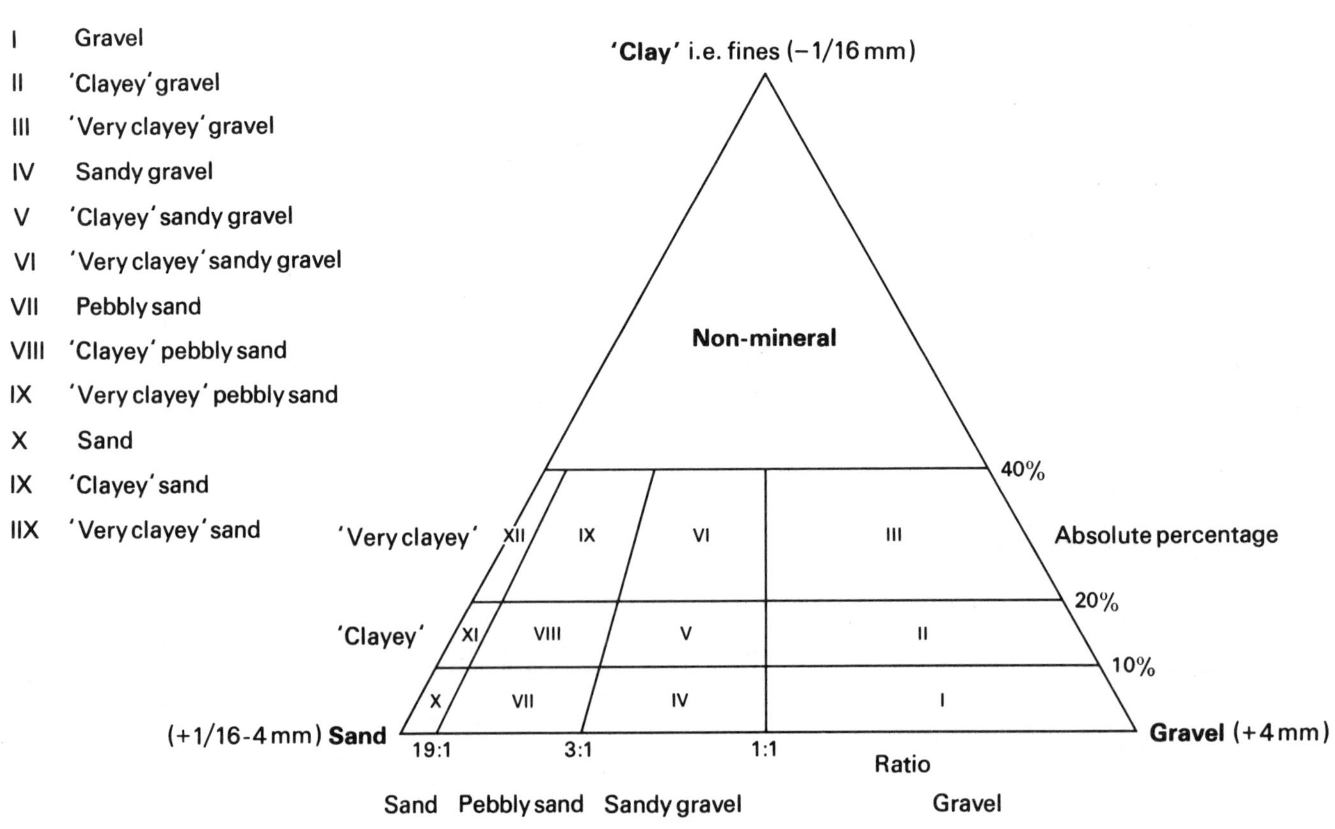

Figure 6 Diagram to show the descriptive categories used in the classification of sand and gravel

APPENDIX D

EXPLANATION OF THE BOREHOLE RECORDS

Annotated Example

SU 88 NE 18[1]   8590 8573[2]   **Quarry Wood, Marlow**[3]   **Block A**

Surface level (+27.5 m) +90 ft[4]   [7]Overburden 1.8 m (6.0 ft)
Water struck at +25.6 m[5]   Mineral 5.6 m (18.5 ft)
Shell and auger (modified) 6-in (152-mm) diameter[6]   Bedrock 0.5 m+ (1.5 ft+)[9]
March 1973

Log

Geological classification	Lithology	Thickness m	Thickness ft	Depth[8] m	Depth ft
	Soil	0.3	(1.0)	0.3	(1.0)
Alluvium[10]	Clay, silty, light grey-yellow	1.5	(5.0)	1.8	(6.0)
	Gravel[11]	5.6	(18.5)	7.4	(24.5)
	Gravel: Coarse and fine with some cobbles, subangular to well rounded flint, well rounded quartz and quartzite with some sandstone and traces of limestone and chalk. Sand: coarse with medium and traces of fine, mainly flint and quartz				
Middle Chalk	Chalk with flint	0.5+	(1.5+)	7.9	(26.0)

Grading[13]

Mean for deposit[14] percentages						Depth below surface (m)	Bulk samples percentages		
Fines	Sand				Gravel		Fines	Sand	Gravel
$-\frac{1}{16}$	$-\frac{1}{4} +\frac{1}{16}$	$-1 +\frac{1}{4}$	$-4 +1$		$-16 +4 +16$				
4	1	9	13		36	37	1.8–2.8[12]	16	20	64
4	23				73		2.8–3.8	0	18	82
							3.8–4.8	2	28	70
							4.8–5.8	1	18	81
							5.8–6.8	1	24	75
							6.8–7.4	4	35	61

The numbered paragraphs below correspond with the annotations given on the specimen record above.

1   Borehole Registration Number
Each Industrial Minerals Assessment Unit (IMAU) borehole is identified by a Registration Number. This consists of two statements.
  1   The number of the 1:25 000 sheet on which the borehole lies, for example SU 88.
  2   The quarter of the 1:25 000 sheet on which the borehole lies and its number in a series for that quarter, for example NE 18.
Thus the full Registration Number is SU 88 NE 18. Usually this is abbreviated to 88 NE 18 in the text.

2   The National Grid reference
All National Grid references in this publication lie within the 100-km square SU unless otherwise stated. Grid references are given to eight figures, accurate to within 10 m for borehole locations. (In the text, six-figure grid references are used for more approximate locations, for example, for farms.)

3   Location
The position of the borehole is generally referred to the nearest named locality on the 1:25 000 base map and the resource block in which it lies is stated.

4   Surface level
The surface level at the borehole site is given in metres and feet above Ordnance Datum. All measurements were made in feet; approximate conversions to metres are given in brackets.

5   Groundwater conditions
If groundwater was present, the level at which it was encountered is normally given (in metres and feet above Ordnance Datum).

6   Type of drill and date of drilling
Modified shell and auger rigs were used in this survey. The type of machine, the external diameter of the casing used, and the month and year of the completion of the borehole are stated.

7   Overburden, mineral, waste and bedrock
Mineral is sand and gravel which, as part of a deposit, falls within the arbitrary definition of potentially workable material (see p. 1). Bedrock is the 'formation', 'country rock' or 'rock head' below which potentially workable sand and gravel will not be found.
Waste is any material other than bedrock or mineral. Where waste occurs between the surface and mineral it is classified as overburden.

8   Thickness and depth
Although most measurements were made in feet, some were recorded in metres; the conversions appear in brackets. Metric conversions, the thicknesses of beds and the depth from the surface of their bases have been rounded off to the nearest 0.1 m because quotation to two places of decimals would imply a higher order of accuracy than could be justified by the original figures. Similarly conversions from metres to feet have been rounded off to the nearest 0.5 ft. Where figures have been rounded in this way there may be a discrepancy between the sum of the thicknesses and the recorded depths.

9   The plus sign ($+$) indicates that the base of the deposit was not reached during drilling.

10   Geological Classification
The geological classification (Table 1) is given whenever possible.

11   Lithological description
When sand and gravel is recorded a general description based on the mean grading characteristics (for details see Appendix C) is followed by more detailed particulars. The description of other rocks is based on visual examination, in the field.

12   Sampling
A continuous series of bulk samples is taken throughout the thickness of sand and gravel. A new sample is commenced whenever there is an appreciable lithological change within the sand and gravel or at every 3 ft or 1 m of depth.

13   Grading results
The limits are as follows: gravel, $+4$ mm; sand, $-4 + \frac{1}{16}$ mm; fines, $-\frac{1}{16}$ mm. If, exceptionally, grading results are not available, an attempt may be made to give grading information by comparing the grading and field descriptions of adjacent samples with the samples in question. Such estimates are shown *in italics*.

14   Mean grading
The grading of the full thickness of the mineral horizon identified in the log is the mean of the individual sample gradings weighted by the thicknesses represented, if these vary. The classification used is shown in Table 9.

Fully representative sampling of sand and gravel is difficult to achieve particularly where groundwater levels are high. Comparison between boreholes and adjacent exposures suggests that in borehole samples the proportion of sand may be higher and the proportions of fines and coarse gravel ($+16$ mm) may be lower.

**LIST OF BOREHOLES USED IN THE
ASSESSMENT OF RESOURCES**

Borehole number*	Grid reference	Borehole number*	Grid reference	Borehole number*	Grid reference	Borehole number*	Grid reference
IMAU BOREHOLES		88 NE 19	8735 8966	97 NW 60	9063 7952	98 NW 70	9029 8540
87 NE 32	8786 7860	88 NE 20	8751 8763	97 NW 61	9060 7881	98 NW 71	9152 8858
87 NE 33	8862 7961	88 NE 21	8764 8714	97 NW 62	9040 7812		
87 NE 34	8984 7938	88 NE 22	8853 8837	97 NW 63	9147 7956	98 SW 67	9025 8360
		88 NE 23	8890 8698	97 NW 64	9154 7774	98 SW 68	9062 8065
88 NW 10	8050 8918	88 NE 24	8796 8631	97 NW 65	9285 7939	98 SW 69	9140 8165
88 NW 11	8065 8524	88 NE 25	8843 8542	97 NW 66	9287 7835	98 SW 70	9168 8052
88 NW 12	8111 8805	88 NE 26	8910 8974	97 NW 67	9372 7889	98 SW 71	9448 8009
88 NW 13	8107 8632	88 NE 27	8961 8920	97 NW 68	9363 7761		
88 NW 14	8279 8711	88 NE 28	8976 8621	97 NW 69	9445 7781		
88 NW 15	8230 8539			97 NW 70	9481 7715		
88 NW 16	8375 8664	88 SW 18	8035 8425				
88 NW 17	8313 8581	88 SW 19	8103 8323	97 NE 187	9556 7960		
88 NW 18	8453 8797	88 SW 20	8151 8396	97 NE 188	9549 7886		
88 NW 19	8447 8557	88 SW 21	8245 8348	97 NE 189	9544 7759		
		88 SW 22	8345 8478				
88 NE 14	8536 8869	88 SW 23	8399 8393	98 NW 66	9037 8944		
88 NE 15	8518 8531			98 NW 67	9033 8857		
88 NE 16	8621 8940	88 SE 26	8804 8388	98 NW 68	9058 8765		
88 NE 17	8655 8803	88 SE 27	8896 8371	98 NW 69	9008 8668		
88 NE 18	8590 8573						

OTHER BOREHOLES

IGS registered boreholes 255/4, 255/216, 269/16, 269/17, 269/139 and 269/278(a).

Many records, which are held in confidence, were made available by the industry.

* By sheet quadrant

APPENDIX F

INDUSTRIAL MINERALS ASSESSMENT UNIT
BOREHOLE RECORDS

SU 87 NE 32 8786 7860 Ockwell's Farm, White Waltham   **Block**

Surface level ($+25.3$ m) $+83$ ft   Waste 1.8 m (6.0 ft)
Water struck at $+24.3$ m   Bedrock 1.2 m+ (4.0 ft+)
Shell and auger (modified) 6-in (152-mm) diameter
April 1973

Log

Geological classification	*Lithology*	*Thickness*		*Depth*	
		m	ft	m	ft
	Soil and made ground	0.3	(1.0)	0.3	(1.0)
Alluvium	Clay, brown to grey-brown with some pebbles of flint	0.8	(2.5)	1.1	(3.5)
	Pebbly sand, fine with coarse gravel composed of subangular to well rounded flint with some quartz and quartzite. Sand, medium with coarse and fine.	0.7	(2.5)	1.8	(6.0)
Reading Beds	Clay, blue-black at top, passes down into mottled red-brown clay.	1.2+	(4.0+)	3.0	(10.0)

SU 87 NE 33 8862 7960 Golf Course, Maidenhead   **Block E**

Surface level ($+44.9$ m) $+147$ ft   Overburden 0.2 m (0.5 ft)
Water not struck   Mineral 5.2 m (17.0 ft)
Shell and auger (modified) 6-in (152-mm) diameter   Bedrock 3.0 m+ (10.0 ft+)
April 1973

Log

Geological classification	*Lithology*	*Thickness*		*Depth*	
		m	ft	m	ft
	Soil	0.2	(0.5)	0.2	(0.5)
River Terrace Deposits (Boyn Hill Terrace)	Gravel	5.2	(17.0)	5.4	(17.5)
	Gravel: fine and coarse with some cobbles of flint. Composed of subangular to well rounded flint with well-rounded quartz and quartzite. Sand: medium with coarse and some fine, brownish yellow.				
Reading Beds	Sandy clay, mottled.	3.0+	(10.0+)	8.4	(27.5)

Grading

Mean for deposit *percentages*						Depth below surface (m)	Bulk samples *percentages*		
Fines	Sand			Gravel			Fines	Sand	Gravel
$-\frac{1}{16}$	$-\frac{1}{4}+\frac{1}{16}$	$-1+\frac{1}{4}$	$-4+1$	$-16+4$	$+16$				
8	3	21	10	29	29	0.2–1.2	6	27	67
						1.2–2.2	6	34	60
8	34			58		2.2–3.2	11	44	45
						3.2–4.2	15	31	54
						4.2–5.4	2	34	64

SU 87 NE 34   **8984 7938**   **Cannon Hill, Bray**   **Block C**

Surface level (+22.9 m) +75 ft   Overburden 1.5 m (5.0 ft)
Water struck at +20.1 m   Mineral 5.2 m (17.0 ft)
Shell and auger (modified) 6-in (152-mm) diameter   Bedrock 2.7 m+ (8.5 ft+)
May 1973

Log

Geological classification	Lithology	Thickness		Depth	
		m	ft	m	ft
	Soil	0.4	(1.5)	0.4	(1.5)
Alluvium	Clay, light brown to grey, sandy and silty with occasional pebbles	1.1	(3.5)	1.5	(5.0)

	Sandy gravel
	    Gravel: coarse and fine with some cobbles. Composed of subangular to well rounded flint with quartz and quartzite
	    Sand: medium with some coarse and fine, light brown to dark brown
		5.2	(17.0)	6.7	(22.0)

Reading Beds	Clay, silty, mottled grey-green	2.2	(7.0)	8.9	(29.0)
Upper Chalk	Chalk with flints	0.5+	(1.5+)	9.4	(30.5)

Grading

Mean for deposit percentages						Depth below surface (m)	Bulk samples percentages		
Fines	Sand			Gravel			Fines	Sand	Gravel
$-\frac{1}{16}$	$-\frac{1}{4}+\frac{1}{16}$	$-1+\frac{1}{4}$	$-4+1$	$-16+4$	$+16$				
7	6	36	6	20	25	1.5–2.4	3	83	14
						2.4–3.4	8	85	7
7	48			45		3.4–4.4	8	40	52
						4.4–5.4	8	14	78
						5.4–6.7	8	28	64

SU 88 NW 10   **8050 8918**   **Chrisbridge Farm, Medmenham**   **Block ...**

Surface level (+152.9 m) +502 ft   Waste 4.0 m (13.0 ft)
Water not struck   Bedrock 0.7 m+ (2.5 ft+)
Shell and auger (modified) 6-in (152-mm) diameter
February 1973

Log

Geological classification	Lithology	Thickness		Depth	
		m	ft	m	ft
	Soil	0.3	(1.0)	0.3	(1.0)
Pebble Gravel	Clay, brown with subangular to subrounded flint and subrounded to rounded pebbles of quartz and quartzite	3.7	(12.0)	4.0	(13.0)
Upper Chalk	Chalk	0.7+	(2.5+)	4.7	(15.5)

SU 88 NW 11   8065 8524   Cobble Wood, Medmenham   **Block F**

Surface level (+91.4 m) +300 ft
Water not struck
Shell and auger (modified) 6-in (152-mm) diameter
February 1973

Overburden 0.2 m (0.5 ft)
Mineral 7.0 m (23.0 ft)
Bedrock 0.5 m+ (1.5 ft+)

Log

Geological classification	Lithology	Thickness		Depth	
		m	ft	m	ft
	Soil	0.2	(0.5)	0.2	(0.5)
Glacial Sand and Gravel	'Very clayey' gravel	7.0	(23.0)	7.2	(23.5)
	Gravel: coarse and fine, well rounded quartz and quartzite with subangular to subrounded flint				
	'Very clayey' sand: medium with coarse and a trace of fine, clayey throughout, brown				
Upper Chalk	Chalk with flints	0.5+	(1.5+)	7.7	(25.0)

Grading

Mean for deposit						Depth below surface (m)	Bulk samples		
percentages							*percentages*		
Fines	Sand				Gravel		Fines	Sand	Gravel
$-\frac{1}{16}$	$-\frac{1}{4}+\frac{1}{16}$	$-1+\frac{1}{4}$	$-4+1$		$-16+4$	$+16$			
21	1	15	9		20	34			
							0.2–1.2	11	26	63
							1.2–2.1	9	35	56
21	25				54		2.1–3.1	15	36	49
							3.1–4.0	30	19	51
							4.0–5.0	30	19	51
							5.0–6.0	26	30	44
							6.0–7.2	23	12	65

SU 88 NW 12   8111 8805   Woodend House, Medmenham   **Block**

Surface level (+145.6 m) +476 ft
Water not struck
Shell and auger (modified) 6-in (152-mm) diameter
February 1973

Waste 5.4 m (17.5 ft)
Bedrock 0.2 m+ (0.5 ft+)

Log

Geological classification	Lithology	Thickness		Depth	
		m	ft	m	ft
	Soil	0.4	(1.5)	0.4	(1.5)
Clay-with-Flints	Clay, yellow-brown with some sandy patches. Contains subrounded to rounded flint, with well rounded quartz, and at the base chalk fragments	5.0	(16.5)	5.4	(17.5)
Upper Chalk	Chalk, with flints	0.2+	(0.5+)	5.6	(18.0)

SU 88 NW 13   8107 8632   Bockmer End, Medmenham   **Block F**

Surface level (+102.6 m) +337 ft
Water not struck
Shell and auger (modified) 6-in (152-mm) diameter
February 1973

Overburden 0.3 m (1.0 ft)
Mineral 5.0 m (16.5 ft)
Waste 0.8 m (2.5 ft)
Bedrock 0.7 m+ (2.5 ft+)

Log

Geological classification	Lithology	Thickness		Depth	
		m	ft	m	ft
	Soil	0.3	(1.0)	0.3	(1.0)
Glacial Sand and Gravel	'Very clayey' pebbly sand	5.0	(16.5)	5.3	(17.5)
	Gravel: coarse and fine, subangular to subrounded flint with well rounded quartz, quartzite and rare sandstone				
	'Very clayey' sand: medium with coarse and some fine, clayey throughout				
	Clay, with cobbles of flint and some chalk	0.8	(2.5)	6.1	(20.0)
Upper Chalk	Chalk with flints	0.7+	(2.5+)	6.8	(22.5)

Grading

Mean for deposit _percentages_						Depth below surface (m)	Bulk samples _percentages_		
Fines	Sand				Gravel		Fines	Sand	Gravel
$-\frac{1}{16}$	$-\frac{1}{4}+\frac{1}{16}$	$-1+\frac{1}{4}$	$-4+1$		$-16+4$	$+16$			
20	13	49	4	5	9	0.3–1.4	18	54	28
						1.4–2.4	24	72	4
20	66			14		2.4–3.4	17	77	6
						3.4–4.4	20	76	4
						4.4–5.3	23	47	30

SU 88 NW 14   8279 8711   Kiln Cottage, Marlow Common   **Block F**

Surface level (+95.2 m) +311 ft
Water not struck
Shell and auger (modified) 6-in (152-mm) diameter
February 1973

Waste 13.5 m (44.5 ft)
Bedrock 0.5 m+ (1.5 ft+)

Log

Geological classification	Lithology	Thickness		Depth	
		m	ft	m	ft
	Soil	0.1	(0.5)	0.1	(0.5)
Glacial Sand and Gravel	Sandy clay, brown with black staining, alternations of sandy clay and 'clayey' sand with subangular to subrounded flints, quartz, stiltsone and quartzite.	13.4	(44.0)	13.5	(44.5)
Upper Chalk	Chalk	0.5+	(1.5+)	14.0	(46.0)

Surface level (+88.6 m) +290 ft
Water not struck
Shell and auger (modified) 6-in (152-mm) diameter
February 1973

Overburden 0.1 m (0.5 ft)
Mineral 3.1 m (10.0 ft)
Waste 1.5 m (5.0 ft)
Mineral 3.7 m (12.0 ft)
Waste 0.6 m (2.0 ft)
Bedrock 0.5 m+ (1.5 ft+)

Log

Geological classification	Lithology	Thickness m	Thickness ft	Depth m	Depth ft
	Soil	0.1	(0.5)	0.1	(0.5)
Glacial Sand and Gravel	**a** 'Very clayey' sandy gravel Gravel: coarse and fine with scattered cobbles, subangular to subrounded flint, quartz, quartzite and sandstone 'Very clayey' sand: medium with coarse and some fine, orange-brown with clay throughout	3.1	(10.0)	3.2	(10.5)
	Clay, orange-brown to light brown	1.5	(5.0)	4.7	(15.5)
	b 'Clayey' sandy gravel Gravel: coarse and fine, well rounded flint, quartzite, quartz and sandstone, scattered flint cobbles and chalk fragments at the base 'Clayey' sand: medium with some coarse and fine orange-brown to light brown, clayey throughout	3.7	(12.0)	8.4	(27.5)
	Clay, orange brown with chalk fragments	0.6	(2.0)	9.0	(29.5)
Upper Chalk	Chalk	0.5+	(1.5+)	9.5	(31.0)

Grading

	Mean for deposit percentages						Depth below surface (m)	Bulk samples percentages		
	Fines	Sand				Gravel		Fines	Sand	Gravel
	$-\frac{1}{16}$	$-\frac{1}{4}+\frac{1}{16}$	$-1+\frac{1}{4}$	$-4+1$	$-16+4$	$+16$				
a	25	4	32	8	13	18	0.1–1.1	24	30	46
							1.1–2.1	14	52	34
	25	44			31		2.1–3.2	38	48	14
b	16	6	40	7	15	16	4.7–5.7	18	49	33
							5.7–6.7	14	60	26
	16	53			31		6.7–7.7	19	52	29
							7.7–8.4	12	50	38

SU 88 NW 16   8375 8664   Forty Green, Marlow   **Block F**

Surface level (+86.2 m) +282 ft
Water not struck
Shell and auger (modified) 6-in (152-mm) diameter
February 1973

Overburden 1.9 m (6.5 ft)
Mineral 1.6 m (5.0 ft)
Bedrock 1.0 m+ (3.5 ft+)

Log

Geological classification	Lithology	Thickness m	Thickness ft	Depth m	Depth ft
	Soil	0.3	(1.0)	0.3	(1.0)
Glacial Sand and Gravel	Sandy clay, brown, fine and coarse subangular to subrounded flint and quartz	1.6	(5.0)	1.9	(6.0)
	'Very clayey' sandy gravel Gravel: coarse with some fine, subangular to subrounded flint with some quartz. Scattered cobbles of flint throughout, chalk framents towards the base 'Very clayey' sand: medium with fine and some coarse, clayey throughout	1.6	(5.0)	3.5	(11.0)
Upper Chalk	Chalk	1.0+	(3.5+)	4.5	(14.5)

Grading

Mean for deposit percentages						Depth below surface (m)	Bulk Samples percentages		
Fines	Sand				Gravel		Fines	Sand	Gravel
$-\frac{1}{16}$	$-\frac{1}{4}+\frac{1}{16}$	$-1+\frac{1}{4}$	$-4+1$		$-16+4$	$+16$			
34	9	30	4		5	18	19.–3.0	36	55	9
34	43				23		3.0–3.5	31	15	54

SU 88 NW 17   8313 8581   Hooks Corner, Marlow   **Block F**

Surface level (+82.8 m) +271 ft
Water not struck
Shell and auger (modified) 6-in (152-mm) diameter
February 1973

Overburden 0.2 m (0.5 ft)
Mineral 3.8 m (12.5 ft)
Waste 2.1 m (7.0 ft)
Bedrock 0.5 m+ (1.5 ft+)

Log

Geological classification	Lithology	Thickness m	Thickness ft	Depth m	Depth ft
	Soil, clayey with fragments of quartz, quartzite and flint	0.2	(0.5)	0.2	(0.5)
Glacial Sand and Gravel	'Clayey' gravel Gravel: coarse and fine with some cobbles, composed of subangular to subrounded flint, and well rounded quartz and quartzite 'Clayey' sand: medium with coarse and fine, brown to red-brown with clay throughout	3.8	(12.5)	4.0	(13.0)
	Clay, brown with fine to coarse angular to subangular flints, quartz, quartzite, and chalk fragments at the base	2.1	(7.0)	6.1	(20.0)
Upper Chalk	Chalk	0.5+	(1.5+)	6.6	(21.5)

Grading

Mean for deposit percentages						Depth below surface (m)	Bulk samples percentages		
Fines	Sand				Gravel		Fines	Sand	Gravel
$-\frac{1}{16}$	$-\frac{1}{4}+\frac{1}{16}$	$-1+\frac{1}{4}$	$-4+1$		$-16+4$	$+16$			
17	3	15	7		23	35	0.2–0.8	34	20	46
17	25				58		0.8–1.8	11	28	61
							1.8–2.7	14	28	58
							2.7–4.0	17	23	60

Surface level (+84.6 m) +277 ft
Water not struck
Shell and auger (modified) 6-in (152-mm) diameter
February 1973

Overburden 0.3 m (1.0 ft)
Mineral 1.5 m (5.0 ft)
Waste 2.7 m (9.0 ft)
Mineral 3.7 m (12.0 ft)
Bedrock 0.5 m+ (1.5 ft+)

Log

Geological classification	Lithology	Thickness		Depth	
		m	ft	m	ft
	Soil	0.3	(1.0)	0.3	(1.0)
Glacial Sand and Gravel	**a** 'Clayey' gravel Gravel: fine with coarse, angular to subangular flint with some well rounded quartz and quartzite 'Clayey' sand: medium and coarse with a trace of fine, brown to red with clay throughout	1.5	(5.0)	1.8	(6.0)
	Sand clay, fine to red-brown with occasional flint pebbles	2.7	(9.0)	4.5	(15.0)
	b 'Very clayey' sandy gravel Gravel: coarse and fine, composed predominantly of angular to subangular flint with minor amounts of well rounded quartz and quartzite, scattered flint cobbles 'Very clayey' sand: medium with coarse and fine, brown to red-brown	3.7	(12.0)	8.2	(27.0)
Upper Chalk	Chalk	0.5+	(1.5+)	8.7	(28.5)

Grading

	Mean for deposit *percentages*						Depth below surface (m)	Bulk samples *percentages*		
	Fines	Sand				Gravel		Fines	Sand	Gravel
	$-\frac{1}{16}$	$-\frac{1}{4} +\frac{1}{16}$	$-1 +\frac{1}{4}$	$-4 +1$		$-16 +4 +16$				
a	15	2	18	14		34	17	0.3–1.4	21	27	52
								1.4–1.8	0	50	50
	15	34				51					
b	20	3	29	9		18	21	4.5–5.5	17	40	43
								5.5–6.5	26	44	30
	20	41				39		6.5–7.5	17	41	42
								7.5–8.2	22	37	41

SU 88 NW 19   8447 8557   South west of Higginson Park, Marlow   **Block A**

Surface level (+27.8 m) +91 ft
Water struck at +26.3 m
Shell and auger (modified) 6-in (152-mm) diameter
February 1973

Overburden 1.5 m (5.0 ft)
Mineral 5.3 m (17.5 ft)
Bedrock 0.5 m+ (1.5 ft+)

Log

Geological classification	Lithology	Thickness		Depth	
		m	ft	m	ft
	Soil	0.2	(0.5)	0.2	(0.5)
Alluvium	Clay, light brown to brown, sandy with some pebbles	1.3	(4.5)	1.5	(5.0)
	'Clayey' gravel	5.3	(17.5)	6.8	(22.5)
	Gravel: coarse with fine and some cobbles. Composed of well rounded flint, quartz and quartzite with some limestone, sandstone, shell debris and chalk				
	'Clayey' sand: coarse and medium with some fine, white, composed of quartz and flint, clayey throughout				
Middle Chalk	Chalk	0.5+	(1.5+)	7.3	(24.0)

Grading

Mean for deposit *percentages*						Depth below surface (m)	Bulk samples *percentages*		
Fines	Sand				Gravel		Fines	Sand	Gravel
$-\frac{1}{16}$	$-\frac{1}{4}+\frac{1}{16}$	$-1+\frac{1}{4}$	$-4+1$		$-16+4$	$+16$			
10	2	16	16	24		32	1.5–2.5	6	47	47
							2.5–3.5	3	33	64
10	34			56			3.5–4.5	35	28	37
							4.5–5.5	5	15	80
							5.5–6.2	2	36	62
							6.2–6.8	5	52	43

SU 88 NE 14   8536 8869   Bencombe Farm, Marlow Bottom   **Block F**

Surface level (+75.3m) +247 ft
Water not struck
Shell and auger (modified) 6-in (152-mm) diameter
March 1973

Overburden 0.3 m (1.0 ft)
Mineral 5.4 m (17.5 ft)
Bedrock 0.5 m+ (1.5 ft+)

Log

Geological classification	Lithology	Thickness		Depth	
		m	ft	m	ft
	Soil	0.3	(1.0)	0.3	(1.0)
Glacial Sand and Gravel	'Clayey' gravel	5.4	(17.5)	5.7	(18.5)
	Gravel: coarse with fine, composed of well rounded flint, quartz and quartzite, cobbles of flint and quartzite				
	'Clayey' sand: medium and coarse with fine, brown, clayey throughout				
Upper Chalk	Chalk	0.5+	(1.5+)	6.2	(20.0)

Grading

Mean for deposit *percentages*						Depth below surface (m)	Bulk samples *percentages*		
Fines	Sand				Gravel		Fines	Sand	Gravel
$-\frac{1}{16}$	$-\frac{1}{4}+\frac{1}{16}$	$-1+\frac{1}{4}$	$-4+1$		$-16+4$	$+16$			
19	3	10	7	19		42	0.3–1.3	5	30	65
							1.3–2.3	12	26	62
19	20			61			2.3–3.3	11	19	70
							3.3–3.8	10	41	49
							3.8–5.7	36	8	56

Surface level (+29.0 m) +95 ft
Water struck at +25.0 m
Shell and auger (modified) 6-in (152-mm) diameter
March 1973

Overburden 1.5 m (5.0 ft)
Mineral 7.6 m (25.0 ft)
Bedrock 0.5 m+ (1.5 ft+)

Log

Geological classification	Lithology	Thickness m	ft	Depth m	ft
	Soil	0.3	(1.0)	0.3	(1.0)
River Terrace Deposits (Flood Plain Terrace)	Sandy clay, brown and white with some pebbles	1.2	(4.0)	1.5	(5.0)
	Gravel	7.6	(25.0)	9.1	(30.0)
	Gravel: coarse with fine, subangular to rounded flint with quartz, quartzite and traces of limestone, sandstone and chalk. Cobbles of flint scattered throughout Sand: medium and coarse with a trace of fine, mainly flint and quartz				
Middle Chalk	Chalk	0.5+	(1.5+)	9.6	(31.5)

Grading

Mean for deposit *percentages*						Depth below suface (m)	Bulk samples *percentages*		
Fines	Sand				Gravel		Fines	Sand	Gravel
$-\frac{1}{16}$	$-\frac{1}{4}+\frac{1}{16}$	$-1+\frac{1}{4}$	$-4+1$		$-16+4+16$				
2	1	18	15		25 39	1.5–2.5	5	44	51
						2.5–3.5	3	53	44
2	34				64	3.5–4.5	3	29	68
						4.5–5.5	0	22	78
						5.5–6.5	1	24	75
						6.5–7.5	0	39	61
						7.5–8.5	1	34	65
						8.5–9.1	0	32	68

Surface level (+91.4 m) +300 ft
Water not struck
Shell and auger (modified) 6-in (152-mm) diameter
March 1973

Overburden 2.2 m (7.0 ft)
Mineral 11.5 m (37.5 ft)
Bedrock 0.6 m+ (2.0 ft+)

Log

Geological classification	Lithology	Thickness		Depth	
		m	ft	m	ft
	Soil	0.3	(1.0)	0.3	(1.0)
Glacial Sand and Gravel	Sandy clay with flints	1.9	(6.0)	2.2	(7.0)
	'Very clayey' gravel	11.5	(37.5)	13.7	(44.5)
	Gravel: coarse with fine and cobbles. Composed of well rounded flint with quartz and quartzite 'Very clayey' sand: medium with coarse and some fine, brown, clayey throughout				
Upper Chalk	Chalk with flint	0.6+	(2.0+)	14.3	(46.5)

Grading

Mean for deposit *percentages*						Depth below surface (m)	Bulk samples *percentages*		
Fines	Sand				Gravel		Fines	Sand	Gravel
$-\frac{1}{16}$	$-\frac{1}{4}+\frac{1}{16}$	$-1+\frac{1}{4}$	$-4+1$		$-16+4$	$+16$			
26	2	11	7		19	35	2.2–3.2	35	19	46
26	20				54		3.2–4.2	38	17	45
							4.2–5.2	16	20	64
							5.2–6.2	33	23	44
							6.2–7.2	23	23	54
							7.2–8.2	23	25	52
							8.2–9.2	21	21	58
							9.2–10.2	28	24	48
							10.2–11.2	10	15	75
							11.2–12.2	14	21	65
							12.2–13.2	29	23	48
							13.2–13.7	39	14	47

SU 88 NE 17   8655 8803   Near Pump Cottage, Little Marlow   **Block E**

Surface level (+40.8 m) +134 ft
Water not struck
Shell and auger (modified) 6-in (152-mm) diameter
March 1973

Overburden 0.3 m (1.0 ft)
Mineral 4.2 m (14.0 ft)
Waste 2.1 m (7.0 ft)
Mineral 3.0 m (10.0 ft)
Bedrock 0.5 m+ (1.5 ft+)

Log

Geological classification	Lithology	Thickness m	ft	Depth m	ft
	Soil	0.3	(1.0)	0.3	(1.0)
River Terrace Deposits (Taplow Terrace)	**a** 'Clayey' gravel Gravel: fine and coarse, subangular to rounded flint with some well rounded quartz and quartzite and traces of chalk. Cobbles of flint scattered throughout 'Clayey' sand: medium and coarse with some fine brown to light brown with clay throughout	4.2	(14.0)	4.5	(15.0)
	Sandy clay, brown	2.1	(7.0)	6.6	(22.0)
	b 'Clayey' gravel Gravel: fine and coarse, subangular to rounded flint with some quartz and quartzite and traces of chalk. Flint cobbles scattered throughout 'Clayey' sand: medium and coarse with a trace of fine, brown to light brown with clay throughout	3.0	(10.0)	9.6	(32.0)
Middle Chalk	Chalk with flint	0.5+	(1.5+)	10.1	(33.5)

Grading

	Mean for deposit percentages						Depth below surface (m)	Bulk samples percentages		
	Fines	Sand				Gravel		Fines	Sand	Gravel
	$-\frac{1}{16}$	$-\frac{1}{4}+\frac{1}{16}$	$-1+\frac{1}{4}$	$-4+1$		$-16+4$	$+16$			
a	13	3	19	10		32	23	0.3–0.9	36	13	51
								0.9–1.9	14	27	59
	13	32				55		1.9–2.9	7	30	63
								2.9–3.9	2	50	48
								3.9–4.5	19	31	50
b	17	1	21	11		28	22	6.6–7.1	39	48	13
								7.1–8.1	11	31	58
	17	33				50		8.1–9.1	9	28	63
								9.1–9.6	20	32	48

SU 88 NE 18   8590 8573   Quarry Wood, Marlow   **Block A**

Surface level (+27.5 m) +90 ft

Water struck at +25.6 m

Shell and auger (modified) 6-in (152-mm) diameter

March 1973

Overburden 1.8 m (6.0 ft)

Mineral 5.6 m (18.5 ft)

Bedrock 0.5 m+ (1.5 ft+)

Log

Geological classification	Lithology	Thickness		Depth	
		m	ft	m	ft
	Soil	0.3	(1.0)	0.3	(1.0)
Alluvium	Clay, silty, light grey-yellow	1.5	(5.0)	1.8	(6.0)
	Gravel:	5.6	(18.5)	7.4	(24.5)
	Gravel: coarse and fine with some cobbles, subangular to well rounded flint, well rounded quartz and quartzite with some sandstone and traces of limestone and chalk				
	Sand: coarse with medium and traces of fine, mainly flint and quartz				
Middle Chalk	Chalk with flint	0.5+	(1.5+)	7.9	(26.0)

Grading

Mean for deposit *percentages*						Depth below surface (m)	Bulk samples *percentages*		
Fines	Sand				Gravel		Fines	Sand	Gravel
$-\frac{1}{16}$	$-\frac{1}{4}+\frac{1}{16}$	$-1+\frac{1}{4}$	$-4+1$		$-16+4$	$+16$			
4	1	9	13	36	37	1.8–2.8	16	20	64
						2.8–3.8	0	18	82
4	23			73		3.8–4.8	2	28	70
						4.8–5.8	1	18	81
						5.8–6.8	1	24	75
						6.8–7.4	4	35	61

SU 88 NE 19   8735 8966   Bloom Wood, Little Marlow   **Block F**

Surface level (+112.3 m) +368 ft

Water not struck

Shell and auger (modified) 6-in (152-mm) diameter

March 1973

Overburden 0.1 m (0.5 ft)

Mineral 4.5 m (15.0 ft)

Bedrock 0.8 m+ (2.5 ft+)

Log

Geological classification	Lithology	Thickness		Depth	
		m	ft	m	ft
	Soil	0.1	(0.5)	0.1	(0.5)
Glacial Sand and Gravel	'Clayey' gravel	4.5	(14.5)	4.6	(15.0)
	Gravel: coarse with fine and with cobbles. Subangular to well rounded flint, quartz and quartzite with some sandstone				
	'Clayey' sand: medium with coarse and with some fine, brown, clayey throughout				
Upper Chalk	Chalk	0.8+	(2.5+)	5.4	(17.5)

Grading

Mean for deposit *percentages*						Depth below surface (m)	Bulk samples *percentages*		
Fines	Sand				Gravel		Fines	Sand	Gravel
$-\frac{1}{16}$	$-\frac{1}{4}+\frac{1}{16}$	$-1+\frac{1}{4}$	$-4+1$		$-16+4$	$+16$			
13	3	16	8	22	38	0.1–1.1	13	30	57
						1.1–2.1	13	25	62
13	27			60		2.1–3.1	12	27	61
						3.1–4.1	12	29	59
						4.1–4.6	15	24	61

SU 88 NE 20   8751 8763   Near St John the Baptists Church, Little Marlow   **Block B**

Surface level (+27.5 m) +90 ft
Water struck at +25 m
Shell and auger (modified) 6-in (152-mm) diameter
March 1973

Overburden 1.6 m (5.0 ft)
Mineral 6.6 m (21.5 ft)
Bedrock 0.5 m+ (1.5 ft+)

Log

Geological classification	Lithology	Thickness m	ft	Depth m	ft
	Soil	0.1	(0.5)	0.1	(0.5)
River Terrace Deposits (Flood Plain Terrace)	Clay: sandy with occasional flint, light brown to dark brown	1.5	(5.0)	1.6	(5.5)
	Gravel	6.6	(21.5)	8.2	(27.0)
	Gravel: coarse with fine and scattered cobbles, subrounded to rounded flint with rounded quartz and quartzite, traces of chalk at base				
	Sand: medium and coarse with traces of fine, brown to light brown				
Middle Chalk	Chalk	0.5+	(1.5+)	8.7	(28.5)

Grading

Mean for deposit percentages						Depth below surface (m)	Bulk samples percentages		
Fines	Sand				Gravel		Fines	Sand	Gravel
$-\frac{1}{16}$	$-\frac{1}{4}+\frac{1}{16}$	$-1+\frac{1}{4}$	$-4+1$		$-16+4$	$+16$			
2	1	13	10		29	45	1.6–2.6	3	25	72
							2.6–3.6	3	28	68
2	24				74		3.6–4.6	2	27	71
							4.6–5.6	3	24	73
							5.6–6.6	0	19	81
							6.6–7.6	1	21	78
							7.6–8.2	1	26	73

SU 88 NE 21   8764 8714   Near Noah's House, Little Marlow   **Block B**

Surface level (+26.8 m) +88 ft
Water struck at +24.6 m
Shell and auger (modified) 6-in (152-mm) diameter
March 1973

Overburden 1.3 m (4.5 ft)
Mineral 5.6 m (18.5 ft)
Bedrock 1.0 m+ (3.5 ft+)

Log

Geological classification	Lithology	Thickness m	ft	Depth m	ft
	Made ground, concrete and soil	1.1	(3.5)	1.1	(3.5)
	Soil	0.2	(0.5)	1.3	(4.0)
River Terrace Deposits (Flood Plain Terrace)	Gravel:	5.6	(18.5)	6.9	(22.5)
	Gravel: coarse and fine with cobbles, sub rounded to rounded flint, quartz and quartzite, with traces of chalk and sandstone				
	Sand: medium with coarse and with some fine, light brown				
Middle Chalk	Chalk	1.0+	(3.5+)	7.9	(26.0)

Grading

Mean for deposit percentages						Depth below surface (m)	Bulk samples percentages		
Fines	Sand				Gravel		Fines	Sand	Gravel
$-\frac{1}{16}$	$-\frac{1}{4}+\frac{1}{16}$	$-1+\frac{1}{4}$	$-4+1$		$-16+4$	$+16$			
9	2	15	9		30	35	1.3–1.8	13	30	57
							1.8–2.8	3	36	61
9	26				65		2.8–3.8	36	16	48
							3.8–4.8	1	25	74
							4.8–5.8	0	31	69
							5.8–6.9	1	16	83

Surface level (+36.5 m) +120 ft	Overburden 0.4 m (1.5 ft)
Water not struck	Mineral 4.6 m (15.0 ft)
Shell and auger (modified) 6-in (152-mm) diameter	Waste 1.9 m (6.0 ft)
March 1973	Mineral 2.2 m (7.0 ft)
	Bedrock 0.5 m+ (1.5 ft+)

Log

Geological classification	Lithology	Thickness		Depth	
		m	ft	m	ft
	Soil	0.4	(1.5)	0.4	(1.5)
River Terrace Deposits (Taplow Terrace)	**a** 'Clayey' gravel Gravel: fine and coarse, well rounded flint, quartz and quartzite with flint cobbles and traces of sandstone 'Clayey' sand: medium with coarse and fine, yellow-brown with clay throughout	4.6	(15.0)	5.0	(16.5)
	Sandy clay, yellow-brown	1.9	(6.0)	6.9	(22.5)
	b 'Clayey' gravel Gravel: fine and coarse, well rounded flint, quartz and quartzite with traces of sandstone, and chalk fragments towards the base. Cobbles of flint scattered throughout 'Clayey' sand: medium with coarse and some fine, yellow-brown with clay throughout	2.2	(7.0)	9.1	(29.5)
Middle Chalk	Chalk	0.5+	(1.5+)	9.6	(31.0)

Grading

	Mean for deposit _percentages_						Depth below surface (m)	Bulk samples _percentages_		
	Fines	Sand				Gravel		Fines	Sand	Gravel
	$-\frac{1}{16}$	$-\frac{1}{4}+\frac{1}{16}$	$-1+\frac{1}{4}$	$-4+1$		$-16+4$	$+16$			
a	18	3	15	9		28	27	0.4–1.2	0	19	81
								1.2–2.2	16	22	62
	18	27				55		2.2–2.7	28	30	42
								2.7–3.2	81	14	5
								3.2–4.2	10	36	54
								4.2–5.0	3	37	60
b	11	3	17	10		25	34	6.9–7.9	10	17	73
								7.9–9.1	12	41	47
	11	30				59					

SU 88 NE 23   **8890 8698**   **Cockmarsh, Cookham**   **Block B**

Surface level (+29.0 m) +95 ft   Overburden 0.5 m (1.5 ft)
Water struck at +28.0 m   Mineral 8.0 m (26.0 ft)
Shell and auger (modified) 6-in (152-mm) diameter   Waste 0.6 m (2.0 ft)
April 1973   Bedrock 0.5 m+ (1.5 ft+)

Log

Geological classification	Lithology	Thickness		Depth	
		m	ft	m	ft
	Soil	0.2	(0.5)	0.2	(0.5)
Alluvium	Clay, dark brown with scattered flint pebbles	0.3	(1.0)	0.5	(1.5)
	Gravel	8.0	(26.0)	8.5	(26.5)
	Gravel: fine and coarse with cobbles, subangular to rounded flint and well rounded quartz and quartzite with traces of sandstone, limestone and chalk Sand: medium and coarse with some fine				
	Sandy clay with chalk fragments	0.6	(2.0)	9.1	(30.0)
Middle Chalk	Chalk	0.5+	(1.5+)	9.6	(31.5)

Grading

Mean for deposit *percentages*						Depth below surface (m)	Bulk samples *percentages*		
Fines	Sand				Gravel		Fines	Sand	Gravel
$-\frac{1}{16}$	$-\frac{1}{4}+\frac{1}{16}$	$-1+\frac{1}{4}$	$-4+1$	$-16+4$	$+16$				
2	2	19	13	29	35	0.5–1.5	3	56	41
						1.5–2.5	1	35	64
2	34			64		2.5–3.5	1	19	80
						3.5–4.5	2	22	76
						4.5–5.5	2	30	68
						5.5–6.5	0	22	78
						6.5–7.5	2	43	55
						7.5–8.5	5	41	54

SU 88 NE 24   **8796 8631**   **Near Long Copse, Cookham**   **Block F**

Surface level (+75.5 m) +248 ft   Overburden 0.3m (1.0 ft)
Water not struck   Mineral 5.9 m (19.5 ft)
Shell and auger (modified) 6-in (152-mm) diameter   Bedrock 0.5 m+ (1.5 ft+)
April 1973

Log

Geological classification	Lithology	Thickness		Depth	
		m	ft	m	ft
	Soil	0.3	(1.0)	0.3	(1.0)
Glacial Sand and Gravel	'Clayey' gravel	5.9	(19.5)	6.2	(20.5)
	Gravel: fine and coarse with cobbles, angular to subrounded flint with some well rounded quartz and quartzite 'Clayey' sand: medium with coarse and some fine, light brown clayey throughout				
Upper Chalk	Chalk	0.5+	(1.5+)	6.7	(22.0)

Grading

Mean for deposit *percentages*						Depth below surface (m)	Bulk samples *percentages*		
Fines	Sand				Gravel		Fines	Sand	Gravel
$-\frac{1}{16}$	$-\frac{1}{4}+\frac{1}{16}$	$-1+\frac{1}{4}$	$-4+1$	$-16+4$	$+16$				
14	4	18	10	29	25	0.3–1.5	8	28	64
						1.5–2.2	30	15	55
14	32			54		2.2–2.6	53	45	2
						2.6–3.6	15	39	46
						3.6–4.6	6	36	58
						4.6–5.6	9	32	59
						5.6–6.2	6	31	63

SU 88 NE 25   **8843 8542**   **Near Cookham Rise, Cookham**   **Block E**

Surface level (+43.5 m) +142 ft
Water not struck
Shell and auger (modified) 6-in (152-mm) diameter
April 1978

Overburden 0.2 m (0.5 ft)
Mineral 3.7 m (12.0 ft)
Bedrock 0.8 m+ (2.5 ft+)

Log

Geological classification	Lithology	Thickness		Depth	
		m	ft	m	ft
	Soil	0.2	(0.5)	0.2	(0.5)
River Terrace Deposits (Boyn Hill Terrace)	'Clayey' gravel	3.7	(12.0)	3.9	(12.5)
	Gravel: coarse with fine and some cobbles, subangular to subrounded flint with well rounded quartz and quartzite				
	'Clayey' sand: medium with coarse and fine, mainly flint and quartz, light brown with clay throughout				
Upper Chalk	Chalk	0.8+	(2.5+)	4.7	(15.0)

Grading

Mean for deposit _percentages_						Depth below surface (m)	Bulk samples _percentages_		
Fines	Sand				Gravel		Fines	Sand	Gravel
$-\frac{1}{16}$	$-\frac{1}{4}+\frac{1}{16}$	$-1+\frac{1}{4}$	$-4+1$		$-16+4$	$+16$			
19	4	15	7		21	34	0.2–1.2	14	18	68
							1.2–1.8	17	17	66
19	26				55		1.8–2.6	25	31	44
							2.6–3.9	21	31	48

SU 88 NE 26   8910 8974   **Chapel Road, Flackwell Heath**   **Block F**

Surface level (+108.5 m) +356 ft
Water not struck
Shell and auger (modified) 6-in (152-mm) diameter
March 1973

Overburden 0.4 m (1.5 ft+)
Mineral 16.3 m (53.5 ft)
Bedrock 0.8 m+ (2.5 ft)

Log

Geological classification	Lithology	Thickness		Depth	
		m	ft	m	ft
	Made ground	0.4	(1.5)	0.4	(1.5)
Glacial Sand and Gravel	'Very clayey' sandy gravel	16.3	(53.5)	16.7	(55.0)
	Gravel: coarse with fine and numerous cobbles. Composed of cobbles and coarse angular to well rounded flint, with fine well rounded quartz and quartzite				
	'Very clayey' sand: medium with fine and coarse, brown, clayey throughout				
Upper Chalk	Chalk	0.8+	(2.5+)	17.5	(57.5)

Grading

Mean for deposit percentages						Depth below surface (m)	Bulk samples percentages		
Fines	Sand				Gravel		Fines	Sand	Gravel
$-\frac{1}{16}$	$-\frac{1}{4}+\frac{1}{16}$	$-1+\frac{1}{4}$	$-4+1$		$-16+4$	$+16$			
21	8	30	5		13	23	0.4–1.2	29	23	48
21	43				36		1.2–2.2	17	45	38
							2.2–3.2	9	38	53
							3.2–3.8	14	33	53
							3.8–4.3	20	79	1
							4.3–5.4	19	66	15
							5.4–6.3	13	66	21
							6.3–7.3	12	75	13
							7.3–8.3	31	57	12
							8.3–9.3	16	41	43
							9.3–10.3	6	19	75
							10.3–11.3	37	14	49
							11.3–11.8	37	18	45
							11.8–12.8	15	30	55
							12.8–13.8	18	53	29
							13.8–14.7	31	36	33
							14.7–15.6	26	46	28
							15.6–16.0	50	13	37
							16.0–16.7	33	49	18

SU 88 NE 27   8961 8920   **Sedgemoor, Flackwell Heath**   **Block F**

Surface level (+109.1 m) +358 ft
Water not struck
Shell and auger (modified) 6-in (152-mm) diameter
March 1973

Waste 5.9 m (19.5 ft)
Bedrock 0.2 m+ (0.5 ft+)

Log

Geological classification	Lithology	Thickness		Depth	
		m	ft	m	ft
	Soil	0.8	(2.5)	0.8	(2.5)
Glacial Sand and Gravel	Sandy clay, brown at top with cobbles of flint, passes down into brown and green clays with flints, and at the base chalk fragments	5.1	(17.0)	5.9	(19.5)
Upper Chalk	Chalk with flints	0.2+	(0.5+)	6.1	(20.0)

SU 88 NE 28   **8976 8621**   **Riversdale Cottages, Bourne End**   **Block B**

Surface level (+25.3 m) +83 ft
Water struck at +23.3 m
Shell and auger (modified) 6-in (152-mm) diameter
March 1973

Overburden 0.7 m (2.5 ft)
Mineral 7.2 m (23.5 ft)
Bedrock 0.5 m+ (1.5 ft+)

Log

Geological classification	Lithology	Thickness		Depth	
		m	ft	m	ft
	Soil	0.2	(0.5)	0.2	(0.5)
Alluvium	Clay, brown, sandy with flint pebbles	0.5	(1.5)	0.7	(2.0)
	'Clayey' gravel	7.2	(23.5)	7.9	(25.5)
	Gravel: fine and coarse, subangular to well rounded flint with well rounded quartz and quartzite, traces of limestone. Cobbles of flint throughout				
	'Clayey' sand: medium and coarse with a trace of fine, white to light brown, clayey in the upper parts				
Upper Chalk	Chalk	0.5+	(1.5+)	8.4	(27.0)

Grading

Mean for deposit _percentages_						Depth below surface (m)	Bulk samples _percentages_		
Fines	Sand			Gravel			Fines	Sand	Gravel
$-\frac{1}{16}$	$-\frac{1}{4} +\frac{1}{16}$	$-1 +\frac{1}{4}$	$-4 +1$	$-16 +4$	$+16$				
11	1	17	16	30	25	0.7–1.7	20	50	30
						1.7–2.7	15	30	55
11	34			55		2.7–3.7	17	22	61
						3.7–4.7	22	38	40
						4.7–5.7	1	28	71
						5.7–6.7	1	31	68
						6.7–7.4	1	34	65
						7.4–7.9	7	39	54

SU 88 SW 18   8035 8425   Manor House, Medmenham   **Block A**

Surface level (+31.4 m) +103 ft
Water struck at + 30.0 m
Shell and auger (modified) 6-in (152-mm) diameter
January 1973

Overburden 1.4 m (4.5 ft)
Mineral 9.9 m (32.5 ft)
Bedrock 0.7 m+ (2.5 ft+)

Log

Geological classification	Lithology	Thickness m	ft	Depth m	ft
	Soil	0.1	(0.5)	0.1	(0.5)
Alluvium	Clay, mottled grey-brown with black staining	1.3	(4.5)	1.4	(5.0)
	Gravel	9.9	(32.5)	11.3	(37.5)
	Gravel: fine with coarse, subrounded to rounded flint, quartz and quartzite, with some chalk and sandstone and traces of limestone and shell debris				
	Sand: coarse and medium with some fine, clean, white to grey				
Middle Chalk	Chalk	0.7+	(2.0+)	12.0	(39.5)

Grading

Mean for deposit *percentages*						Depth below surface (m)	Bulk samples *percentages*		
Fines	Sand			Gravel			Fines	Sand	Gravel
$-\frac{1}{16}$	$-\frac{1}{4}+\frac{1}{16}$	$-1+\frac{1}{4}$	$-4+1$	$-16+4$	$+16$				
7	2	15	18	37	21	1.4–2.4	11	31	58
						2.4–3.4	4	42	54
7	35			58		3.4–4.4	3	45	52
						4.4–5.4	3	26	71
						5.4–6.4	3	44	53
						6.4–7.4	0	25	75
						7.4–8.4	1	25	74
						8.4–9.4	9	26	65
						9.4–10.2	12	26	62
						10.2–11.3	27	59	14

SU 88 SW 19   8103 8323   Frogmill Farm, Hurley   **Block A**

Surface level (+29.9 m) +98 ft
Water struck at +28.2 m
Shell and auger (modified) 6-in (152-mm) diameter
February 1973

Overburden 1.5 m (5.0 ft)
Mineral 4.9 m (16.0 ft)
Bedrock 0.6 m+ (2.0 ft+)

Log

Geological classification	Lithology	Thickness m	ft	Depth m	ft
	Soil	0.2	(0.5)	0.2	(0.5)
River Terrace Deposits (Flood Plain Terrace)	Clay, light brown to brown with flint cobbles	1.3	(4.5)	1.5	(5.0)
	Gravel	4.9	(16.0)	6.4	(21.0)
	Gravel: fine and coarse with cobbles, angular to well rounded flint with well rounded quartz and quartzite, and trace amounts of sandstone, limestone and chalk				
	Sand: medium with coarse and fine, composed of flint and quartz				
Upper Chalk	Chalk	0.6+	(2.0+)	7.0	(23.0)

Grading

Mean for deposit *percentages*						Depth below surface (m)	Bulk samples *percentages*		
Fines	Sand			Gravel			Fines	Sand	Gravel
$-\frac{1}{16}$	$-\frac{1}{4}+\frac{1}{16}$	$-1+\frac{1}{4}$	$-4+1$	$-16+4$	$+16$				
5	2	16	6	36	35	1.5–2.6	4	34	62
						2.6–3.5	4	38	58
5	24			71		3.5–4.5	0	31	69
						4.5–5.5	14	27	59
						5.5–6.4	1	25	74

Surface level (+29.5 m) +97 ft
Water struck at +27.3 m
Shell and auger (modified) 6-in (152-mm) diameter
February 1973

Overburden 2.2 m (7.0 ft)
Mineral 6.7 m (22.0 ft)
Bedrock 0.5 m+ (1.5 ft+)

Log

Geological classification	Lithology	Thickness		Depth	
		m	ft	m	ft
	Soil	0.2	(0.5)	0.2	(0.5)
Alluvium	Silt and clay, brown at the top becoming blue-grey towards the base with occasional shell fragments	2.0	(6.5)	2.2	(7.0)
	Gravel	6.7	(22.0)	8.9	(29.0)
	Gravel: coarse and fine with cobbles. Composed of subangular to subrounded flint with well rounded quartz and quartzite, and traces of chalk, limestone, sandstone and shell debris				
	Sand: medium and coarse with a trace of fine, white, composed of flint and quartz				
Upper Chalk	Chalk	0.5+	(1.5+)	9.4	(30.5)

Grading

Mean for deposit *percentages*						Depth below surface (m)	Bulk samples *percentages*		
Fines	Sand				Gravel		Fines	Sand	Gravel
$-\frac{1}{16}$	$-\frac{1}{4} +\frac{1}{16}$	$-1 +\frac{1}{4}$	$-4 +1$		$-16 +4$	$+16$			
2	1	10	9		34	44	2.2–3.2	6	6	88
2	20				78		3.2–4.2	0	21	79
							4.2–5.2	2	17	81
							5.2–6.2	1	11	88
							6.2–7.2	0	31	69
							7.2–8.2	3	31	66
							8.2–8.9	1	27	72

SU 88 SW 21   8245 8348   Lee Farm House, Hurley Bottom   **Block A**

Surface level (+28.3 m) +93 ft
Water struck at +25.3 m
Shell and auger (modified) 6-in (152-mm) diameter
February 1973

Overburden 1.0 m (3.5 ft)
Mineral 5.6 m (18.5 ft)
Bedrock 0.5 m+ (1.5 ft+)

Log

Geological classification	Lithology	Thickness		Depth	
		m	ft	m	ft
	Soil	0.3	(1.0)	0.3	(1.0)
River Terrace Deposits (Flood Plain Terrace)	Clay, sandy, dark brown with some pebbles	0.7	(2.5)	1.0	(3.5)
	Gravel	5.6	(18.5)	6.6	(21.5)
	Gravel: fine and coarse with cobbles. Composed of subangular to subrounded flint with quartz, quartzite and some sandstone, limestone and chalk				
	Sand: coarse and medium with a trace of fine, clean white, composed of quartz and flint				
Upper Chalk	Chalk	0.5+	(1.5+)	7.1	(23.5)

Grading

Mean for deposit percentages						Depth below surface (m)	Bulk samples percentages		
Fines	Sand			Gravel			Fines	Sand	Gravel
$-\frac{1}{16}$	$-\frac{1}{4}+\frac{1}{16}$	$-1+\frac{1}{4}$	$-4+1$	$-16+4$	$+16$				
5	3	16	16	32	28	1.0–1.4	24	30	46
						1.4–2.4	8	29	63
5	35			60		2.4–3.4	4	32	64
						3.4–4.4	2	37	61
						4.4–5.4	1	40	59
						5.4–6.6	3	38	59

SU 88 SW 22   8345 8478   Low Grounds, Marlow   **Block A**

Surface level (+29.0 m) +95 ft
Water struck at +27.7 m
Shell and auger (modified) 6-in (152-mm) diameter
February 1973

Overburden 1.3 m (4.5 ft)
Mineral 5.8 m (19.0 ft)
Bedrock 0.5 m+ (1.5 ft+)

Log

Geological classification	Lithology	Thickness		Depth	
		m	ft	m	ft
	Soil	0.2	(0.5)	0.2	(0.5)
River Terrace Deposits (Flood Plain Terrace)	Clay: silty-sandy, light brown to greyish brown with small pebbles of flint	1.1	(3.5)	1.3	(4.0)
	Gravel	5.8	(19.0)	7.1	(23.5)
	Gravel: fine and coarse with cobbles. Composed of subangular to subrounded flint with subrounded to well rounded quartz, quartzite and sandstone, and some chalk and limestone				
	Sand: medium and coarse with some fine, white to pale brown composed of flint and quartz				
Upper Chalk	Chalk	0.5+	(1.5+)	7.6	(25.0)

Grading

Mean for deposit percentages						Depth below surface (m)	Bulk samples percentages		
Fines	Sand			Gravel			Fines	Sand	Gravel
$-\frac{1}{16}$	$-\frac{1}{4}+\frac{1}{16}$	$-1+\frac{1}{4}$	$-4+1$	$-16+4$	$+16$				
3	2	18	15	30	32	1.3–2.3	3	42	55
						2.3–3.3	2	34	64
3	35			62		3.3–4.3	0	27	73
						4.3–5.3	3	29	68
						5.3–6.3	1	27	72
						6.3–7.1	8	56	36

Surface level ($+28.3$ m) $+93$ ft   Overburden 0.8 m (2.5 ft)
Water struck at $+26.4$ m   Mineral 5.5 m (18.0 ft)
Shell and auger (modified) 6-in (152-mm) diameter   Bedrock 0.5 m+ (1.5 ft+)
January 1973

Log

Geological classification	Lithology	Thickness m	ft	Depth m	ft
	Soil	0.1	(0.5)	0.1	(0.1)
River Terrace Deposits (Flood Plain Terrace)	Clay, brown to light brown, silty	0.7	(2.5)	0.8	(2.5)
	Gravel	5.5	(18.0)	6.3	(20.5)
	Gravel: fine and coarse with cobbles. Composed of subangular to subrounded flint and well rounded quartz with some quartzite and chalk Sand: medium and coarse with a trace of fine, white to pale brown, composed mainly of flint and quartz				
Upper Chalk	Chalk with flints	0.5+	(1.5+)	6.8	(22.5)

Grading

Mean for deposit percentages						Depth below surface (m)	Bulk samples percentages		
Fines	Sand			Gravel			Fines	Sand	Gravel
$-\frac{1}{16}$	$-\frac{1}{4}+\frac{1}{16}$	$-1+\frac{1}{4}$	$-4+1$	$-16+4$	$+16$				
7	1	23	15	28	26	0.8–1.2	14	20	66
						1.2–2.2	3	32	65
7	39			54		2.2–3.2	15	56	29
						3.2–4.2	5	44	51
						4.2–5.2	1	35	64
						5.2–5.8	4	40	56
						5.8–6.3	10	35	55

Surface level ($+41.0$ m) $+134$ ft   Overburden 0.8 m (2.5 ft)
Water not struck   Mineral 2.1 m (7.0 ft)
Shell and auger (modified) 6-in (152-mm) diameter   Bedrock 0.5 m+ (1.5 ft+)
April 1973

Log

Geological classification	Lithology	Thickness m	ft	Depth m	ft
	Made ground, old road	0.8	(2.5)	0.8	(2.5)
River Terrace Deposits (Boyn Hill Terrace)	Gravel	2.1	(7.0)	2.9	(9.5)
	Gravel: fine and coarse with some cobbles. Composed of angular to subangular flint with well rounded quartz and quartzite and traces of sandstone. Chalk fragments at base Sand: coarse and medium with a trace of fine				
Upper Chalk	Chalk	0.5+	(1.5+)	3.4	(11.0)

Grading

Mean for deposit percentages						Depth below surface (m)	Bulk samples percentages		
Fines	Sand			Gravel			Fines	Sand	Gravel
$-\frac{1}{16}$	$-\frac{1}{4}+\frac{1}{16}$	$-1+\frac{1}{4}$	$-4+1$	$-16+4$	$+16$				
6	2	13	13	33	33	0.8–1.8	6	30	64
						1.8–2.9	6	27	67
6	28			66					

SU 88 SE 27　　**8896 8371**　　**Widbrook Common, Cookham**　　　　　　　　　　**Block B**

Surface level (+24.4 m) +80 ft
Water struck at +20.9 m
Shell and auger (modified) 6-in (152-mm) diameter
April 1973

Overburden 2.2 m (7.0 ft)
Mineral 3.9 m (13.0 ft)
Bedrock 0.5 m+ (1.5 ft+)

Log

Geological classification	Lithology	Thickness		Depth	
		m	ft	m	ft
	Soil	0.3	(1.0)	0.3	(1.0)
Alluvium	Sandy clay, yellow brown	1.9	(6.0)	2.2	(7.0)
	Gravel	3.9	(13.0)	6.1	(20.0)
	Gravel: fine and coarse with cobbles. Composed of subangular to well rounded flint, with quartz, quartzite and traces of limestone and sandstone Sand: coarse and medium with a trace of fine				
Upper Chalk	Chalk	0.5+	(1.5+)	6.6	(21.5)

Grading

Mean for deposit *percentages*						Depth below surface (m)	Bulk samples *percentages*		
Fines	Sand				Gravel		Fines	Sand	Gravel
$-\frac{1}{16}$	$-\frac{1}{4}+\frac{1}{16}$	$-1+\frac{1}{4}$	$-4+1$		-16 $+4$ $+16$				
3	1	10	10		41　35	2.2–3.2	4	28	68
						3.2–4.2	3	23	74
3	21				76	4.2–5.2	1	17	82
						5.2–6.1	5	13	82

SU 97 NW 60　　**9063 7952**　　**River Gardens, Bray**　　　　　　　　　　**Block C**

Surface level (+23.1 m) +76 ft
Water struck at +19.3 m
Shell and auger (modified) 6-in (152-mm) diameter
April 1973

Overburden 0.6 m (2.0 ft)
Mineral 7.8 m (25.5 ft)
Bedrock 0.6 m (2.0 ft)

Log

Geological classification	Lithology	Thickness		Depth	
		m	ft	m	ft
	Soil	0.3	(1.0)	0.3	(1.0)
River Terrace Deposits (Flood Plain Terrace)	Clay, sandy with some pebbles, light brown	0.3	(1.0)	0.6	(2.0)
	Gravel	7.8	(25.5)	8.4	(27.5)
	Gravel: fine and coarse with cobbles. Composed of subangular to well rounded flint with well rounded quartz and quartzite, and traces of chalk and sandstone Sand: medium with coarse and a trace of fine, brown				
Reading Beds	Clay: mottled red, brown and grey, becoming dark grey with depth	0.6+	(2.0+)	9.0	(29.5)

Grading

Mean for deposit *percentages*						Depth below surface (m)	Bulk samples *percentages*		
Fines	Sand				Gravel		Fines	Sand	Gravel
$-\frac{1}{16}$	$-\frac{1}{4}+\frac{1}{16}$	$-1+\frac{1}{4}$	$-4+1$		-16 $+4$ $+16$				
7	3	21	13		29　27	0.3–1.2	37	53	10
						1.2–2.2	6	42	52
7	37				56	2.2–3.2	4	58	38
						3.2–4.2	4	29	67
						4.2–5.2	1	29	70
						5.2–6.2	1	28	71
						6.2–7.2	0	21	79
						7.2–8.1	4	34	62

SU 97 NW 61   9060 7881   The Cut, Bray   **Block C**

Surface level (+21.6 m) +71 ft
Water struck at + 19.1 m
Shell and auger (modified) 6-in (152-mm) diameter
April 1973

Overburden 1.8 m (6.0 ft)
Mineral 6.1 m (20.0 ft)
Bedrock 0.5 m+ (1.5 ft+)

Log

Geological classification	Lithology	Thickness		Depth	
		m	ft	m	ft
	Soil	0.3	(1.0)	0.3	(1.0)
Alluvium	Silty clay, light brown with some flint pebbles	1.5	(5.0)	1.8	(6.0)
	Gravel	6.1	(20.0)	7.9	(26.0)
	Gravel: fine and coarse with cobbles. Composed of subangular to well rounded flint, well rounded quartz and quartzite with traces of sandstone Sand: coarse and medium, blue-grey to light brown				
Reading Beds	Clay, mottled red-brown and grey	0.5+	(1.5+)	8.4	(27.5)

Grading

Mean for deposit percentages						Depth below surface (m)	Bulk samples percentages		
Fines	Sand				Gravel		Fines	Sand	Gravel
$-\frac{1}{16}$	$-\frac{1}{4}+\frac{1}{16}$	$-1+\frac{1}{4}$	$-4+1$		$-16+4$	$+16$			
1	0	9	12		41	37	1.8–2.8	1	31	68
							2.8–3.8	1	15	84
1	21				78		3.8–4.8	1	20	79
							4.8–5.8	0	7	93
							5.8–6.8	2	21	77
							6.8–7.9	1	29	70

SU 97 NW 62   9040 7812   Stroud Form, Bray   **Block E**

Surface level (+24.7 m) +81 ft
Water not struck
Shell and auger (modified) 6-in (152-mm) diameter
May 1973

Overburden 0.9 m (3.0 ft)
Mineral 2.9 m (9.5 ft)
Bedrock 0.5 m+ (1.5 ft+)

Log

Geological classification	Lithology	Thickness		Depth	
		m	ft	m	ft
	Soil	0.4	(1.5)	0.4	(1.5)
River Terrace Deposits (Taplow Terrace)	Clay, sandy, light brown with pebbles of flint	0.5	(1.5)	0.9	(3.0)
	Gravel	2.9	(9.5)	3.8	(12.5)
	Gravel: fine and coarse flint with quartz and quartzite Sand: medium and coarse with some fine, light brown				
Reading Beds	Clay, mottled red, brown and grey	0.5+	(1.5+)	4.3	(14.0)

Grading

Mean for deposit percentages						Depth below surface (m)	Bulk samples percentages		
Fines	Sand				Gravel		Fines	Sand	Gravel
$-\frac{1}{16}$	$-\frac{1}{4}+\frac{1}{16}$	$-1+\frac{1}{4}$	$-4+1$		$-16+4$	$+16$			
9	3	17	13		30	28	0.9–1.9	11	44	45
							1.9–2.9	7	29	64
9	33				58		2.9–3.8	10	25	65

SU 97 NW 63   9147 7956   Dorney Reach, Dorney   **Block C**

Surface level (+22.2 m) +73 ft
Water struck at +19.5 m
Shell and auger (modified) 6-in (152-mm) diameter
May 1973

Overburden 2.4 m (8.0 ft)
Mineral 5.3 m (17.5 ft)
Bedrock 0.5 m+ (1.5 ft+)

Log

Geological classification	Lithology	Thickness m	ft	Depth m	ft
	Soil	0.2	(0.5)	0.2	(0.5)
River Terrace Deposits (Flood Plain Terrace)	Clay, light brown to brown, sandy with rare pebbles	2.2	(7.0)	2.4	(7.5)
	Gravel	5.3	(17.5)	7.7	(25.0)
	Gravel: fine and coarse with some flint cobbles. Composed of subangular to well rounded flint with well rounded quartz, quartzite, some chalk and traces of sandstone Sand: coarse and medium with a trace of fine				
Reading Beds	Clay, mottled red, brown and grey	0.5+	(1.5+)	8.2	(26.5)

Grading

Mean for deposit percentages						Depth below surface (m)	Bulk samples percentages		
Fines	Sand			Gravel			Fines	Sand	Gravel
$-\frac{1}{16}$	$-\frac{1}{4}+\frac{1}{16}$	$-1+\frac{1}{4}$	$-4+1$	$-16+4$	$+16$				
1	1	12	14	38	34	2.4–3.4	2	30	68
						3.4–4.4	3	35	62
1	27			72		4.4–5.4	0	16	84
						5.4–6.4	0	23	77
						6.4–7.7	1	31	68

SU 97 NW 64   9154 7774   Queens Head, Bray   **Block E**

Surface level (+25.0 m) +82 ft
Water struck at +20.2 m
Shell and auger (modified) 6-in (152-mm) diameter
April 1973

Overburden 1.1 m (3.5 ft)
Mineral 5.1 m (17.0 ft)
Bedrock 0.7 m+ (2.5 ft+)

Log

Geological classification	Lithology	Thickness m	ft	Depth m	ft
	Soil	0.4	(1.5)	0.4	(1.5)
River Terrace Deposits (Taplow Terrace)	Clay, brown, sandy with some flint pebbles	0.7	(2.0)	1.1	(3.5)
	'Clayey' gravel	5.1	(17.0)	6.2	(20.5)
	Gravel: fine and coarse with some cobbles. Composed of subangular to well rounded flint with well rounded quartzite and quartz, and some sandstone 'Clayey' sand: medium and coarse with a trace of fine, clayey at the top				
London Clay	Clay, brown to dark brown, silty	0.7+	(2.0+)	6.9	(22.5)

Grading

Mean for deposit percentages						Depth below surface (m)	Bulk samples percentages		
Fines	Sand			Gravel			Fines	Sand	Gravel
$-\frac{1}{16}$	$-\frac{1}{4}+\frac{1}{16}$	$-1+\frac{1}{4}$	$-4+1$	$-16+4$	$+16$				
12	2	14	10	30	32	1.1–2.1	16	27	57
						2.1–3.1	41	20	39
12	26			62		3.1–4.1	4	31	65
						4.1–5.1	0	15	85
						5.1–6.2	1	34	65

SU 97 NW 65   9285 7939   The Vicarage, Dorney   **Block C**

Surface level (+22.5 m) +74 ft
Water struck at +19.2 m
Shell and auger (modified) 6-in (152-mm) diameter
April 1973

Overburden 0.9 m (3.0 ft)
Mineral 4.8 m (15.5 ft)
Bedrock 1.1 m+ (3.5 ft+)

Log

Geological classification	Lithology	Thickness m	ft	Depth m	ft
	Made ground	0.5	(1.5)	0.5	(1.5)
River Terrace Deposits (Flood Plain Terrace)	Clay, brown, sandy with occasional flint pebbles	0.4	(1.5)	0.9	(3.0)
	Gravel	4.8	(15.5)	5.7	(18.5)
	Gravel: coarse and fine with some cobbles. Composed of subangular to well rounded flint, and well rounded quartz, with quartzite and some limestone, sandstone and chalk Sand: medium and coarse with a trace of fine				
Reading Beds	Clay, mottled blue, blue-grey and brown, slightly sandy	1.1+	(3.5+)	6.8	(22.0)

Grading

Mean for deposit _percentages_						Depth below surface (m)	Bulk samples _percentages_		
Fines	Sand				Gravel		Fines	Sand	Gravel
$-\frac{1}{16}$	$-\frac{1}{4}+\frac{1}{16}$	$-1+\frac{1}{4}$	$-4+1$	$-16+4$	$+16$				
6	2	14	12	30	36	0.9–2.0	11	45	44
						2.0–3.0	5	34	61
6	28			66		3.0–4.0	3	11	86
						4.0–5.0	14	28	58
						5.0–6.7	1	24	75

SU 97 NW 66   9287 7835   Near Elm Farm, Dorney   **Block C**

Surface level (+22.3 m) +73 ft
Water struck at +20.5 m
Shell and auger (modified) 6-in (152-mm) diameter
April 1973

Overburden 1.3 m (4.5 ft)
Mineral 5.6 m (18.5 ft)
Bedrock 0.6 m+ (2.0 ft+)

Log

Geological classification	Lithology	Thickness m	ft	Depth m	ft
	Made ground; old road	0.6	(2.0)	0.6	(2.0)
River Terrace Deposits (Flood Plain Terrace)	Clay, mottled brown, grey and black with some pebbles	0.7	(2.5)	1.3	(4.5)
	Gravel	5.6	(18.5)	6.9	(23.0)
	Gravel: coarse and fine with cobbles. Composed of subangular to well rounded flints and well rounded quartz and quartzite with traces of sandstone and chalk Sand: medium and coarse with a trace of fine, white to light brown				
London Clay	Clay, blue-grey	0.6+	(2.0+)	7.5	(25.0)

Grading

Mean for deposit _percentages_						Depth below surface (m)	Bulk samples _percentages_		
Fines	Sand				Gravel		Fines	Sand	Gravel
$-\frac{1}{16}$	$-\frac{1}{4}+\frac{1}{16}$	$-1+\frac{1}{4}$	$-4+1$	$-16+4$	$+16$				
3	1	12	12	37	35	1.3–2.3	9	39	52
						2.3–3.3	0	25	75
3	25			72		3.3–4.3	1	23	76
						4.3–5.3	0	14	86
						5.3–6.3	6	18	76
						6.3–6.9	4	34	62

SU 97 NW 67 **9372 7889** **Manor Farm, Dorney** **Block D**

Surface level (+22.3 m) +73 ft
Water struck at +19.3 m
Shell and auger (modified) 6-in (152-mm) diameter
April 1973

Overburden 1.0 m (3.5 ft)
Mineral 6.9 m (22.5 ft)
Bedrock 1.1 m+ (3.5 ft+)

Log

Geological classification	Lithology	Thickness m	ft	Depth m	ft
	Soil	0.4	(1.5)	0.4	(1.5)
River Terrace Deposits (Flood Plain Terrace)	Clay, brown to light brown with some flint pebbles	0.6	(2.0)	1.0	(3.5)
	Gravel	6.9	(22.5)	7.9	(26.0)
	Gravel: coarse and fine with cobbles. Composed of angular to well rounded flint with well rounded quartz and quartzite, and traces of chalk and sandstone. Sand: medium with coarse and a trace of fine, clayey in upper half, white in colour				
	Clay, blue-grey, silty and sandy at the top	1.1+	(3.5+)	9.0	(29.5)

Grading

Mean for deposit percentages						depth below surface (m)	Bulk samples percentages		
Fines	Sand			Gravel			Fines	Sand	Gravel
$-\frac{1}{16}$	$-\frac{1}{4}+\frac{1}{16}$	$-1+\frac{1}{4}$	$-4+1$	$-16+4$	$+16$				
8	2	18	11	30	31	1.0–2.0	16	42	42
						2.0–3.0	6	57	37
8	31			61		3.0–4.0	8	43	49
						4.0–5.0	16	24	60
						5.0–6.0	6	16	78
						6.0–7.0	1	24	75
						7.0–7.9	3	41	56

SU97 NW 68 **9363 7761** **Boveney Court, Boveney** **Block D**

Surface level +(21.0 m) +69 ft
Water struck at +19.4 m
Shell and auger (modified) 6-in (152-mm) diameter
April 1973

Overburden 1.6 m (5.0 ft)
Mineral 5.9 m (19.5 ft)
Bedrock 0.9 m+ (3.0 ft+)

Log

Geological classification	Lithology	Thickness m	ft	Depth m	ft
	Soil	0.4	(1.0)	0.4	(1.0)
River Terrace Deposits (Flood Plain Terrace)	Clay, light brown with pebbles, sandy at the base	1.2	(4.0)	1.6	(5.0)
	Gravel	5.9	(19.5)	7.5	(24.5)
	Gravel: fine and coarse with cobbles. Composed of angular to well rounded flint, well rounded quartz and quartzite, with some sandstone and traces of limestone. Sand: medium with coarse and some fine, light brown, composed of flint and quartz				
London Clay	Clay, brown and dark grey, silty	0.9+	(3.0+)	8.4	(27.5)

Grading

Mean for deposit percentages						Depth below surface (m)	Bulk samples percentages		
Fines	Sand			Gravel			Fines	Sand	Gravel
$-\frac{1}{16}$	$-\frac{1}{4}+\frac{1}{16}$	$-1+\frac{1}{4}$	$-4+1$	$-16+4$	$+16$				
3	3	18	10	35	31	1.6–2.6	5	33	62
						2.6–3.6	2	37	61
3	31			66		3.6–4.6	2	14	84
						4.6–5.6	0	24	76
						5.6–6.6	1	29	70
						6.6–7.5	10	44	46

SU 97 NW 69   9445 7781   Boveney Lock, Boveney   **Block D**

Surface level (+21.0 m) +69 ft

Water struck at +17.2 m

Shell and auger (modified) 6-in (152-mm) diameter

April 1973

Overburden 1.4 m (4.5 ft)

Mineral 5.1 m (17.0 ft)

Bedrock 0.5 m+ (1.5 ft+)

Log

Geological classification	*Lithology*	*Thickness*		*Depth*	
		m	ft	m	ft
	Made ground	0.1	(0.5)	0.1	(0.5)
River Terrace Deposits (Flood Plain Terrace)	Sandy clay, brown with some flint pebbles	1.3	(4.0)	1.4	(4.5)
	Gravel	5.1	(17.0)	6.5	(21.5)
	Gravel: fine and coarse, angular to well rounded flint with well rounded quartz and quartzite and traces of chalk. Cobbles of flint scattered throughout Sand: medium with coarse and fine, light brown				
London Clay	Clay, blue-grey	0.5+	(1.5+)	7.0	(23.0)

Grading

Mean for deposit *percentages*						Depth below surface (m)	Bulk samples *percentages*		
Fines	Sand				Gravel		Fines	Sand	Gravel
$-\frac{1}{16}$	$-\frac{1}{4}+\frac{1}{16}$	$-1+\frac{1}{4}$	$-4+1$		$-16+4$	$+16$			
7	4	22	9		28	30	1.4–2.4	26	68	6
							2.4–3.4	3	34	63
7	35				58		3.4–4.4	0	26	74
							4.4–5.4	3	25	72
							5.4–6.5	3	22	75

SU 97 NW 70   9481 7715   Windsor Race Course, Windsor   **Block D**

Surface level +(21.0 m) +69 ft

Water struck at 18.7 m

Shell and auger (modified) 6-in (152-mm) diameter

May 1973

Overburden 1.7 m (5.5 ft)

Mineral 4.1.m (13.5 ft)

Bedrock 0.5 m+ (1.5 ft+)

Log

Geological classification	*Lithology*	*Thickness*		*Depth*	
		m	ft	m	ft
	Soil	0.3	(1.0)	0.3	(1.0)
River Terrace Deposits (Flood Plain Terrace)	Clay, light brown, sandy with some flint pebbles	1.4	(4.5)	1.7	(5.5)
	Gravel	4.1	(13.5)	5.8	(19.0)
	Gravel: fine and coarse, subangular to rounded flint with well rounded quartz and quartzite, and cobbles of flint Sand: medium with coarse and some fine				
London Clay	Clay, dark grey	0.5+	(1.5+)	6.3	(20.5)

Grading

Mean for deposit *percentages*						Depth below surface (m)	Bulk samples *percentages*		
Fines	Sand				Gravel		Fines	Sand	Gravel
$-\frac{1}{16}$	$-\frac{1}{4}+\frac{1}{16}$	$-1+\frac{1}{4}$	$-4+1$		$-16+4$	$+16$			
3	4	15	11		34	33	1.7–2.7	6	37	57
							2.7–3.7	2	27	71
3	30				67		3.7–4.7	1	17	82
							4.7–5.8	1	40	59

SU 97 NE 187   9556 7960   Charley Grove, Slough   **Block D**

Surface level (+21.4 m) +70 ft
Water struck at 17.5 m
Shell and auger (modified) 6-in (152-mm) diameter
May 1973

Overburden 1.3 m (4.5 ft)
Mineral 6.5 m (21.5 ft)
Bedrock 0.6 m+ (2.0 ft+)

Log

Geological classification	Lithology	Thickness		Depth	
		m	ft	m	ft
	Soil	0.2	(0.5)	0.2	(0.5)
Alluvium	Clay, sandy with some pebbles	1.1	(4.0)	1.3	(4.5)
	Gravel	6.5	(21.5)	7.8	(26.0)
	Gravel: coarse with fine and some cobbles of flint. Composed of subangular to well rounded flint with well rounded quartz, quartzite and traces of sandstone and limestone				
	Sand: medium with coarse and some fine, white to brown, clayey at top				
Reading Beds	Clay, hard, mottled red and brown	0.6+	(2.0+)	8.4	(28.0)

Grading

Mean for deposit percentages						Depth below surface (m)	Bulk samples percentages		
Fines	Sand				Gravel		Fines	Sand	Gravel
$-\frac{1}{16}$	$-\frac{1}{4}+\frac{1}{16}$	$-1+\frac{1}{4}$	$-4+1$		$-16+4$	$+16$			
4	5	19	10		26	36	1.3–2.4	24	38	38
							2.4–3.4	3	28	69
4	34				62		3.4–4.4	0	36	64
							4.4–5.4	1	36	63
							5.4–6.4	1	36	63
							6.4–7.8	0	26	74

SU 97 NE 188   9549 7886   Manor Farm, Eton   **Block D**

Surface level (+21.0 m) +69 ft
Water struck at +18.0 m
Shell and auger (modified) 6-in (152-mm) diameter
May 1973

Overburden 1.9 m (6.0 ft)
Mineral 4.8 m (15.5 ft)
Bedrock 0.5 m+ (1.5 ft+)

Log

Geological classification	Lithology	Thickness		Depth	
		m	ft	m	ft
	Made ground, old road	0.4	(1.5)	0.4	(1.5)
River Terrace Deposits (Flood Plain Terrace)	Clay, yellow-brown with some pebbles	1.5	(4.5)	1.9	(6.0)
	'Clayey' gravel, contains a thin seam of clay and peat	4.8	(15.5)	6.7	(21.5)
	Gravel: coarse with fine and cobbles of flint. Composed of angular to subrounded flint with well rounded quartz and quartzite				
	'Clayey' sand: medium and coarse with a trace of fine white to pale brown				
London Clay	Clay, dark grey, silty	0.5+	(1.5+)	7.2	(23.0)

Grading

Mean for deposit percentages						Depth below surface (m)	Bulk samples percentages		
Fines	Sand				Gravel		Fines	Sand	Gravel
$-\frac{1}{16}$	$-\frac{1}{4}+\frac{1}{16}$	$-1+\frac{1}{4}$	$-4+1$		$-16+4$	$+16$			
12	1	9	8		25	45	1.9–2.5	28	32	40
							2.5–3.0	Seam of clay over peat		
12	18				70		3.0–4.0	14	14	72
							4.0–5.0	19	12	69
							5.0–6.0	0	25	75
							6.0–6.7	1	9	90

SU 97 NE 189   9544 7759   Parade Ring, Windsor Race Course, Windsor   **Block D**

Surface level (+19.5 m) +64 ft
Water struck at 16.8 m
Shell and auger (modified) 6-in (152-mm) diameter
May 1973

Overburden 2.7 m (9.0 ft)
Mineral 3.2 m (10.5 ft)
Bedrock 0.5 m+ (1.5 ft+)

Log

Geological classification	Lithology	Thickness m	ft	Depth m	ft
	Made ground	0.5	(1.5)	0.5	(1.5)
Alluvium	Clay, grey, silty with scattered pebbles and shells, peaty in middle	2.2	(7.5)	2.7	(9.0)
	Gravel	3.2	(10.5)	5.9	(19.5)
	Gravel: coarse with fine, subangular to rounded flint with well rounded quartz and quartzite, and some shell fragments				
	Sand: medium and coarse, composed of flint and quartz				
London Clay	Clay, blue-grey	0.5+	(1.5+)	6.4	(21.0)

Grading

Mean for deposit percentages						Depth below surface (m)	Bulk samples percentages		
Fines	Sand			Gravel			Fines	Sand	Gravel
$-\frac{1}{16}$	$-\frac{1}{4}+\frac{1}{16}$	$-1+\frac{1}{4}$	$-4+1$	$-16+4$	$+16$				
2	0	14	10	30	44	2.7–3.7	1	9	90
						3.7–4.7	1	17	82
2	24			74		4.7–5.9	3	43	54

SU 98 NW 66   9037 8944   Golf Course, Flackwell Heath   **Block E**

Surface level (+111.1 m) +364 ft
Water not struck
Shell and auger (modified) 6-in (152-mm) diameter
March 1973

Overburden 0.1 m (0.5 ft)
Mineral 5.4 m (17.5 ft)
Bedrock 7.7 m (25.5 ft)

Log

Geological classification	Lithology	Thickness m	ft	Depth m	ft
	Soil	0.1	(0.5)	0.1	(0.5)
Pebble Clay	'Clayey' gravel	5.4	(17.5)	5.5	(18.0)
	Gravel: coarse with fine, subangular to subrounded flint with well rounded quartz and quartzite and a trace of sandstone. Scattered flint cobbles				
	'Clayey' sand: medium with coarse and fine, clayey throughout				
Reading Beds	Clay, mottled grey, brown and red	4.3	(14.0)	9.8	(32.0)
	Sand, clayey, medium grained with some fine and a trace of coarse, white	3.0	(10.0)	12.8	(42.0)
	Clay, dark brown to black with large flints and fragments of chalk	0.3	(1.0)	13.1	(43.0)
Chalk	Chalk	0.1+	(0.5+)	13.2	(43.5)

Grading

Mean for deposit percentages						Depth below surface (m)	Bulk samples percentages		
Fines	Sand			Gravel			Fines	Sand	Gravel
$-\frac{1}{16}$	$-\frac{1}{4}+\frac{1}{16}$	$-1+\frac{1}{4}$	$-4+1$	$-16+4$	$+16$				
15	6	23	7	19	30	0.1–1.0	9	21	70
						1.0–2.0	9	26	65
15	36			49		2.0–2.8	22	76	2
						2.8–3.8	15	36	49
						3.8–4.8	16	35	49
						4.8–5.5	25	31	44

SU 98 NW 67   9033 8857   Ronald Wood, Flackwell Heath   **Block F**

Surface level (+91.4 m) +300 ft
Water not struck
Shell and auger (modified) 6-in (152-mm) diameter
March 1973

Waste 3.6 m (12.0 ft)
Bedrock 0.5 m+ (1.5 ft+)

Log

Geological classification	*Lithology*	*Thickness*		*Depth*	
		m	ft	m	ft
	Soil, clayey with flints	0.4	(1.5)	0.4	(1.5)
Glacial Sand and Gravel	Clay, sandy with fragments of quartz, quartzite and cobbles of flint	3.2	(10.5)	3.6	(12.0)
Upper Chalk	Chalk with flints	0.5+	(1.5+)	4.1	(13.5)

SU 98 NW 68   9058 8765   The Mill, Wooburn   **Block B**

Surface level (+32.6 m) +107 ft
Water struck at +29.1 m
Shell and auger (modified) 6-in (152-mm) diameter
April 1973

Overburden 0.6 m (2.0 ft)
Mineral 5.5 m (18.0 ft)
Bedrock 0.5 m+ (1.5 ft+)

Log

Geological classification	*Lithology*	*Thickness*		*Depth*	
		m	ft	m	ft
	Soil	0.4	(1.5)	0.4	(1.5)
River Terrace Deposits (Flood Plain Terrace)	Clay, light grey, silty with some small pebbles	0.2	(0.5)	0.6	(2.0)
	Gravel	5.5	(18.0)	6.1	(20.0)
	Gravel: coarse with fine, subangular to well rounded flint with quartz and quartzite. Chalk scattered throughout increasing with depth, flint cobbles at base Sand: coarse and medium with a trace of fine, composed of quartz and flint				
Upper Chalk	Chalk	0.5+	(1.5+)	6.6	(21.5)

Grading

Mean for deposit *percentages*						Depth below surface (m)	Bulk samples *percentages*		
Fines	Sand			Gravel			Fines	Sand	Gravel
$-\frac{1}{16}$	$-\frac{1}{4}+\frac{1}{16}$	$-1+\frac{1}{4}$	$-4+1$	$-16+4$	$+16$				
5	2	10	11	31	41	0.6–1.6	8	17	75
						1.6–2.6	7	29	64
5	23			72		2.6–3.6	3	24	73
						3.6–4.6	4	17	79
						4.6–5.6	1	17	82
						5.6–6.1	9	35	56

SU 98 NW 69   9008 8668   Wharton House, Hedsor   **Block B**

Surface level ($+31.4$ m) $+103$ ft
Water struck at $+26$ m
Shell and auger (modified) 6-in (152-mm) diameter
March 1973

Overburden 3.2 m (10.5 ft)
Mineral 5.6 m (18.5 ft)
Bedrock 0.5 m+ (1.5 ft+)

Log

Geological classification	Lithology	Thickness		Depth	
		m	ft	m	ft
	Soil	0.2	(0.5)	0.2	(0.5)
River Terrace Deposits (Flood Plain Terrace)	Clay, dark brown, silty-sandy with some flint pebbles	3.0	(10.0)	3.2	(10.5)
	'Clayey' gravel	5.6	(18.5)	8.8	(29.0)
	Gravel: fine and coarse with scattered cobbles, subangular to well rounded flint, quartz and quartzite with some sandstone, limestone and chalk at the base 'Clayey' sand: coarse and medium with a trace of fine, clayey in the upper parts				
Upper Chalk	Chalk with flints	0.5+	(1.5+)	9.3	(30.5)

Grading

	Mean for deposit *percentages*						Depth below surface (m)	Bulk samples *percentages*		
Fines	Sand				Gravel			Fines	Sand	Gravel
$-\frac{1}{16}$	$-\frac{1}{4}+\frac{1}{16}$	$-1+\frac{1}{4}$	$-4+1$		$-16+4+16$					
13	2	15	14		28	28	3.2–4.2	11	24	65
							4.2–4.9	16	24	60
13	31				56		4.9–5.9	2	34	64
							5.9–6.9	38	18	44
							6.9–7.9	2	31	67
							7.9–8.8	7	57	36

SU 98 NW 70   9029 8540   Odney Common, Cookham   **Block B**

Surface level ($+26.5$ m) $+87$ ft
Water struck at $+24.7$ m
Shell and auger (modified) 6-in (152-mm) diameter
April 1973

Overburden 1.3 m (4.5 ft)
Mineral 5.7 m (18.5 ft)
Bedrock 0.5 m+ (1.5 ft+)

Log

Geological classification	Lithology	Thickness		Depth	
		m	ft	m	ft
	Soil	0.1	(0.5)	0.1	(0.5)
Alluvium	Clay, light brown, silty	1.2	(4.0)	1.3	(4.5)
	Gravel	5.7	(18.5)	7.0	(23.0)
	Gravel: fine and coarse, subangular to subrounded flint with well rounded quartz and quartzite, and traces of chalk and sandstone. Scattered cobbles of flint Sand: coarse and medium with a trace of fine				
Upper Chalk	Chalk	0.5+	(1.5+)	7.5	(24.5)

Grading

	Mean for deposit *percentages*						Depth below surface (m)	Bulk samples *percentages*		
Fines	Sand				Gravel			Fines	Sand	Gravel
$-\frac{1}{16}$	$-\frac{1}{4}+\frac{1}{16}$	$-1+\frac{1}{4}$	$-4+1$		$-16+4+16$					
4	2	8	11		45	30	1.3–2.3	8	30	62
							2.3–3.3	3	10	87
4	21				75		3.3–4.3	0	10	90
							4.3–5.3	4	30	66
							5.3–6.3	8	24	68
							6.3–7.0	0	22	78

SU 98 NW 71   9152 8858   Manor Farm, Wooburn Green   **Block B**

Surface level (+36.9 m) +121 ft
Water struck at +34.4 m
Shell and auger (modified) 6-in (152-mm) diameter
March 1973

Overburden 0.3 m (1.0 ft)
Mineral 4.7 m (15.5 ft)
Bedrock 0.5 m+ (1.5 ft+)

Log

Geological classification	Lithology	Thickness		Depth	
		m	ft	m	ft
	Soil	0.3	(1.0)	0.3	(1.0)
Alluvium	Gravel	4.7	(15.5)	5.0	(16.5)
	Gravel: coarse with fine, subangular to rounded flint with well rounded quartz, quartzite and some chalk. Scattered cobbles of flint Sand: coarse with medium and a trace of fine, composed of quartz and flint				
Upper Chalk	Chalk	0.5+	(1.5+)	5.5	(18.0)

Grading

Mean for deposit *percentages*						Depth below surface (m)	Bulk samples *percentages*		
Fines	Sand				Gravel		Fines	Sand	Gravel
$-\frac{1}{16}$	$-\frac{1}{4}+\frac{1}{16}$	$-1+\frac{1}{4}$	$-4+1$	$-16+4$	$+16$				
4	1	13	14	28	40	0.3–1.3	5	23	72
						1.3–2.3	4	33	63
4	28			68		2.3–3.3	1	20	79
						3.3–4.3	4	33	63
						4.3–5.0	6	30	64

SU 98 SW 67   9025 8360   Maidenhead Court, Maidenhead   **Block B**

Surface level (+25.3 m) +83 ft
Water struck at +22.9 m
Shell and auger (modified) 6-in (152-mm) diameter
May 1973

Overburden 0.8 m (2.5 ft)
Mineral 5.6 m (18.5 ft)
Bedrock 0.6 m+ (2.0 ft+)

Log

Geological classification	Lithology	Thickness		Depth	
		m	ft	m	ft
	Soil	0.8	(2.5)	0.8	(2.5)
River Terrace Deposits (Flood Plain Terrace)	Gravel	5.6	(18.5)	6.4	(21.0)
	Gravel: coarse and fine, subangular to rounded flint with well rounded quartz and quartzite, some chalk and traces of sandstone and limestone. Scattered cobbles of flint Sand: medium with coarse and some fine				
Upper Chalk	Chalk	0.6+	(2.0)	7.0	(23.0)

Grading

Mean for deposit *percentages*						Depth below surface (m)	Bulk samples *percentages*		
Fines	Sand				Gravel		Fines	Sand	Gravel
$-\frac{1}{16}$	$-\frac{1}{4}+\frac{1}{16}$	$-1+\frac{1}{4}$	$-4+1$	$-16+4$	$+16$				
5	3	16	11	33	32	0.8–1.3	27	71	2
						1.3–2.3	7	44	49
5	30			65		2.3–3.3	3	25	72
						3.3–4.3	0	16	84
						4.3–5.3	1	21	78
						5.3–6.4	1	27	72

SU 98 SW 68   9062 8065   Amerden Ponds, Taplow   **Block C**

Surface level (+22.6 m) +74 ft
Water not struck
Shell and auger (modified) 6-in (152-mm) diameter
April 1973

Overburden 2.5 m (8.0 ft)
Mineral 4.4 m (14.5 ft)
Bedrock 0.5 m+ (1.5 ft+)

Log

Geological classification	Lithology	Thickness		Depth	
		m	ft	m	ft
	Soil	0.1	(0.5)	0.1	(0.5)
River Terrace Deposits	Clay, light brown to grey, black at base, silty with some shells	2.1	(6.5)	2.2	(7.0)
(Flood Plain Terrace)	Peat, black and silty	0.3	(1.0)	2.5	(8.0)
	Gravel	4.4	(14.5)	6.9	(22.5)
	Gravel: coarse and fine, subangular to rounded flint with well rounded quartz and quartzite, traces of sandstone, and abundant chalk at the base Sand: coarse and medium with a trace of fine				
Upper Chalk	Chalk	0.5+	(1.5+)	7.4	(24.0)

Grading

Mean for deposit percentages						Depth below surface (m)	Bulk samples percentages		
Fines	Sand						Fines	Sand	Gravel
					Gravel				
$-\frac{1}{16}$	$-\frac{1}{4}+\frac{1}{16}$	$-1+\frac{1}{4}$	$-4+1$		$-16+4+16$				
3	1	10	10	33	43	2.5–3.5	6	14	80
						3.5–4.5	2	11	87
3	21			76		4.5–5.5	1	17	82
						5.5–6.5	0	31	69
						6.5–6.9	5	53	42

SU 98 SW 69   9140 8165   Berry Hill, Taplow   **Block E**

Surface level (+32.4 m) +106 ft
Water not struck
Shell and auger (modified) 6-in (152-mm) diameter
April 1973

Overburden 1.0,m (3.5 ft)
Mineral 5.5 m (8.0 ft)
Bedrock 0.9 m+ (3.0 ft+)

Log

Geological classification	Lithology	Thickness		Depth	
		m	ft	m	ft
	Soil	0.2	(0.5)	0.2	(0.5)
River Terrace Deposits	Clay, brown and sandy with some flint pebbles	0.8	(3.0)	1.0	(3.5)
(Taplow Terrace)	'Clayey' sandy gravel	5.5	(18.0)	6.5	(21.5)
	Gravel: fine with coarse, subangular to subrounded flint with some well rounded quartz and quartzite 'Clayey' sand: medium with coarse and fine, clay throughout				
Upper Chalk	Chalk	0.9+	(3.0+)	7.4	(24.5)

Grading

Mean for deposit percentages						Depth below surface (m)	Bulk samples percentages		
Fines	Sand						Fines	Sand	Gravel
					Gravel				
$-\frac{1}{16}$	$-\frac{1}{4}+\frac{1}{16}$	$-1+\frac{1}{4}$	$-4+1$		$-16+4+16$				
12	5	34	8	26	15	1.0–2.0	22	61	17
						2.0–3.0	14	72	14
12	47			41		3.0–4.0	15	74	11
						4.0–5.0	13	79	8
						5.0–5.5	13	73	14
						5.5–6.1	0	58	42
						6.1–6.5	24	43	33

SU 98 SW 70   9168 8052   Marsh Lane, Taplow   **Block C**

Surface level (+22.9 m) +75 ft
Water struck at 20.2 m
Shell and auger (modified) 6-in (152-mm) diameter
April 1973

Overburden 0.8 m (2.5 ft)
Mineral 4.2 m (14.0 ft)
Bedrock 1.0 m+ (3.5 ft+)

Log

Geological classification	Lithology	Thickness m	ft	Depth m	ft
	Soil	0.2	(0.5)	0.2	(0.5)
River Terrace Deposits	Clay, sandy with some flint pebbles	0.6	(2.0)	0.8	(2.5)
(Flood Plain Terrace)	Gravel	4.2	(14.0)	5.0	(16.5)
	Gravel: fine and coarse, subangular to well rounded flint with well rounded quartz and quartzite, some limestone, sandstone and chalk. Scattered cobbles of flint throughout Sand: medium with coarse and some fine				
Reading Beds	Clay, mottled brown-grey	1.0+	(3.5+)	6.0	(20.0)

Grading

Mean for deposit *percentages*						Depth below surface (m)	Bulk samples *percentages*		
Fines	Sand			Gravel			Fines	Sand	Gravel
$-\frac{1}{16}$	$-\frac{1}{4}+\frac{1}{16}$	$-1+\frac{1}{4}$	$-4+1$	$-16+4$	$+16$				
7	3	23	11	36	20	0.8–1.8	15	43	42
						1.8–2.8	2	31	67
7	37			56		2.8–3.8	5	38	57
						3.8–5.0	5	38	57

SU 98 SW 71   9448 8009   Cippenham Court Farm, Slough   **Block D**

Surface level (+22.9 m) +75 ft
Water struck at +19.3 m
Shell and auger (modified) 6-in (152-mm) diameter
May 1973

Overburden 1.9 m (6.0 ft)
Mineral 5.1 m (17.0 ft)
Bedrock 0.5 m+ (1.5 ft+)

Log

Geological classification	Lithology	Thickness m	ft	Depth m	ft
	Soil	0.3	(1.0)	0.3	(1.0)
River Terrace Deposits	Clay, sandy with some flint pebbles	1.6	(5.0)	1.9	(6.0)
(Flood Plain Terrace)	Gravel	5.1	(17.0)	7.0	(23.0)
	Gravel: coarse and fine, angular to well rounded flint with some well rounded quartz and quartzite Sand: medium and coarse with some fine				
Reading Beds	Clay, mottled brown, blue and grey	0.5+	(1.5+)	7.5	(24.5)

Grading

Mean for deposit *percentages*						Depth below surface (m)	Bulk samples *percentages*		
Fines	Sand			Gravel			Fines	Sand	Gravel
$-\frac{1}{16}$	$-\frac{1}{4}+\frac{1}{16}$	$-1+\frac{1}{4}$	$-4+1$	$-16+4$	$+16$				
4	2	13	14	37	30	1.9–2.9	7	36	57
						2.9–3.9	6	39	55
4	29			67		3.9–4.9	5	24	71
						4.9–5.9	1	16	83
						5.9–7.0	1	30	69

m	ft	m	ft	m	ft	m	ft	m	ft
0.1	0.5	6.1	20	12.1	39.5	18.1	59.5	24.1	79
0.2	0.5	6.2	20.5	12.2	40	18.2	59.5	24.2	79.5
0.3	1	6.3	20.5	12.3	40.5	18.3	60	24.3	79.5
0.4	1.5	6.4	21	12.4	40.5	18.4	60.5	24.4	80
0.5	1.5	6.5	21.5	12.5	41	18.5	60.5	24.5	80.5
0.6	2	6.6	21.5	12.6	41.5	18.6	61	24.6	80.5
0.7	2.5	6.7	22	12.7	41.5	18.7	61.5	24.7	81
0.8	2.5	6.8	22.5	12.8	42	18.8	61.5	24.8	81.5
0.9	3	6.9	22.5	12.9	42.5	18.9	62	24.9	81.5
1.0	3.5	7.0	23	13.0	42.5	19.0	62.5	25.0	82
1.1	3.5	7.1	23.5	13.1	43	19.1	62.5	25.1	82.5
1.2	4	7.2	23.5	13.2	43.5	19.2	63	25.2	82.5
1.3	4.5	7.3	24	13.3	43.5	19.3	63.5	25.3	83
1.4	4.5	7.4	24.5	13.4	44	19.4	63.5	25.4	83.5
1.5	5	7.5	24.5	13.5	44.5	19.5	64	25.5	83.5
1.6	5	7.6	25	13.6	44.5	19.6	64.5	25.6	84
1.7	5.5	7.7	25.5	13.7	45	19.7	64.5	25.7	84.5
1.8	6	7.8	25.5	13.8	45.5	19.8	65	25.8	84.5
1.9	6	7.9	26	13.9	45.5	19.9	65.5	25.9	85
2.0	6.5	8.0	26	14.0	46	20.0	65.5	26.0	85.5
2.1	7	8.1	26.5	14.1	46.5	20.1	66	26.1	85.5
2.2	7	8.2	27	14.2	46.5	20.2	66.5	26.2	86
2.3	7.5	8.3	27	14.3	47	20.3	66.5	26.3	86.5
2.4	8	8.4	27.5	14.4	47	20.4	67	26.4	86.5
2.5	8	8.5	28	14.5	47.5	20.5	67.5	26.5	87
2.6	8.5	8.6	28	14.6	48	20.6	67.5	26.6	87.5
2.7	9	8.7	28.5	14.7	48	20.7	68	26.7	87.5
2.8	9	8.8	29	14.8	48.5	20.8	68	26.8	88
2.9	9.5	8.9	29	14.9	49	20.9	68.5	26.9	88.5
3.0	10	9.0	29.5	15.0	49	21.0	69	27.0	88.5
3.1	10	9.1	30	15.1	49.5	21.1	69	27.1	89
3.2	10.5	9.2	30	15.2	50	21.2	69.5	27.2	89
3.3	11	9.3	30.5	15.3	50	21.3	70	27.3	89.5
3.4	11	9.4	31	15.4	50.5	21.4	70	27.4	90
3.5	11.5	9.5	31	15.5	51	21.5	70.5	27.5	90
3.6	12	9.6	31.5	15.6	51	21.6	71	27.6	90.5
3.7	12	9.7	32 ·	15.7	51.5	21.7	71	27.7	91
3.8	12.5	9.8	32	15.8	52	21.8	71.5	27.8	91
3.9	13	9.9	32.5	15.9	52	21.9	72	27.9	91.5
4.0	13	10.0	33	16.0	52.5	22.0	72	28.0	92
4.1	13.5	10.1	33	16.1	53	22.1	72.5	28.1	92
4.2	14	10.2	33.5	16.2	53	22.2	73	28.2	92.5
4.3	14	10.3	34	16.3	53.5	22.3	73	28.3	93
4.4	14.5	10.4	34	16.4	54	22.4	73.5	28.4	93
4.5	15	10.5	34.5	16.5	54	22.5	74	28.5	93.5
4.6	15	10.6	35	16.6	54.5	22.6	74	28.6	94
4.7	15.5	10.7	35	16.7	55	22.7	74.5	28.7	94
4.8	15.5	10.8	35.5	16.8	55	22.8	75	28.8	94.5
4.9	16	10.9	36	16.9	55.5	22.9	75	28.9	95
5.0	16.5	11.0	36	17.0	56	23.0	75.5	29.0	95
5.1	17	11.1	36.5	17.1	56	23.1	76	29.1	95.5
5.2	17	11.2	36.5	17.2	56.5	23.2	76	29.2	96
5.3	17.5	11.3	37	17.3	57	23.3	76.5	29.3	96
5.4	17.5	11.4	37.5	17.4	57	23.4	77	29.4	96.5
5.5	18	11.5	37.5	17.5	57.5	23.5	77	29.5	97
5.6	18.5	11.6	38	17.6	57.5	23.6	77.5	29.6	97
5.7	18.5	11.7	38.5	17.7	58	23.7	78	29.7	97.5
5.8	19	11.8	38.5	17.8	58.5	23.8	78	29.8	98
5.9	19.5	11.9	39	17.9	58.5	23.9	78.5	29.9	98
6.0	19.5	12.0	39.5	18.0	59	24.0	78.5	30.0	98.5

REFERENCES

ALLEN, V. T. 1936. Terminology of medium-grained sediments. *Rep. Natl. Res. Coun. Washington, 1935–1936. App. 1, Rep. Comm. Sedimentation*, pp. 18–47.

ARCHER, A. A. 1969. Background and problems of an assessment of sand and gravel resources in the United Kingdom. *Proc. 9th Commonw. Min. Metall. Congr. 1969.* Vol. 2: Mining and petroleum geology. (London: Institution of Mining and Metallurgy.) Pp. 495–508.

— 1970 a. Standardisation of the size classification of naturally occurring particles. *Géotechnique*, Vol. 20, pp. 103–207.

— 1970b. Making the most of metrication. *Quarry Manager's J.*, Vol. 54, pp. 223–227.

ATTERBERG, A. 1905. Die rationelle klassification der Sande und Kiese. *Chem. Z.*, Vol. 29, pp. 195–198.

BRITISH STANDARD 1377. 1967. *Methods of testing soils for civil engineering purposes.* (London: British Standards Institution.)

BUREAU OF MINES AND GEOLOGICAL SURVEY. 1948. *Mineral resources of the United States.* (Washington, DC: Public Affairs Press.) pp. 14–17.

DEWEY, H. and BROMEHEAD G. E. N. 1915. The geology of the country around Windsor and Chertsey. *Mem. Geol. Surv. G.B.*

HARE, F. K. 1947. The geomorphology of a part of the Middle Thames. *Proc. Geol. Assoc.*, Vol. 58, pp. 294–339.

HARRIS, P. M., THURRELL, R. G., HEALING, R. A. and ARCHER, A. A. 1974. Aggregates in Britain. *Proc. R. Soc.*, Ser. A, Vol. 339, pp. 329–353.

LANE, E. W. and others. 1947. Report of the subcommittee on sediment terminology. *Trans. Am. Geophys. Union*, Vol. 28, pp. 936–938.

PETTIJOHN, F. J. 1957. *Sedimentary rocks* (2nd edition). (London: Harper and Row.)

SHERLOCK, R. L. and NOBLE, A. H. 1922. The geology of the country around Beaconsfield. *Mem. Geol. Surv. G.B.*

SQUIRRELL, H. C. 1974. The sand and gravel resources of the country around Gerrards Cross, Buckinghamshire: Description of parts of 1:25 000 resource sheets SU 98, SU 99, TQ 08 and TQ 09. *Rep. Inst. Geol. Sci.*, No. 74/14.

THURRELL, R. G. 1971. The assessment of mineral resources with particular reference to sand and gravel. *Quarry Manager's J.*, Vol. 55, pp. 19–25.

TWENHOFEL, W. H. 1937. Terminology of the fine grained mechanical sediments. *Rep. Natl. Res. Coun. Washington 1936–1937. App. 1, Rep. Comm. Sedimentation*, pp. 81–104.

UDDEN, J. A. 1914. Mechanical composition of clastic sediments. *Bull Geol. Soc. Am.*, Vol. 25, pp. 655–744.

WENTWORTH, C. K. 1922. A scale of grade and class terms for clastic sediments. *J. Geol.*, Vol. 30, pp. 377–392.

— 1935. The terminology of coarse sediments. *Bull. Natl Res. Coun. Washington*, No. 98, pp. 225–246.

WHITE, H. J. O. 1899. On the origin of the high level gravel with Triassic debris adjoining the valley of the Upper Thames. *Proc. Geol. Assoc.*, Vol. 15, pp. 157–174.

— 1906. On the occurrence of quartzose gravel in the Reading Beds at Lane End, Bucks. *Proc. Geol. Assoc.*, Vol. 19, pp. 371–379.

WILLMAN, H. B. 1942. Geology and mineral resources of the Marseilles, Ottawa and Streatar quadrangles. *Bull. Illinois. State Geol. Surv.*, No. 66, pp. 343–344.

WOOLRIDGE, S. W. 1938. The glaciation of the London Basin and the evolution of the Lower Thames drainage system. *Q.J. Geol. Soc. London*, Vol. 94, pp. 627–667.

The following reports of the Institute relate particularly to bulk mineral resources

Reports of the Institute of Geological Sciences

Assessment of British Sand and Gravel Resources

1   The sand and gravel resources of the country south-east of Norwich, Norfolk: Resource sheet TG 20.   E. F. P. Nickless.
Report 71/20   ISBN 0 11 880216   £1.15

2   The sand and gravel resources of the country around Witham, Essex: Resource sheet TL 81.   H. J. E. Haggard.
Report 72/6   ISBN 0 11 880588 6   £1.20

3   The sand and gravel resources of the area south and west of Woodbridge, Suffolk: Resource sheet TM 24.
R. Allender and S. E. Hollyer.
Report 72/9   ISBN 0 11 880596 7   £1.70

4   The sand and gravel resources of the country around Maldon, Essex: Resource sheet TL 80.   J. D. Ambrose.
Report 73/1   ISBN 0 11 880600 9   £1.20

5   The sand and gravel resources of the country around Hethersett, Norfolk: Resource sheet TG 10.   E. F. P. Nickless.
Report 73/4   ISBN 0 11 880606 8   £1.60

6   The sand and gravel resources of the country around Terling, Essex: Resource sheet TL 71.   C. H. Eaton.
Report 73/5   ISBN 0 11 880608 4   £1.20

7   The sand and gravel resources of the country around Layer Breton and Tolleshunt D'Arcy, Essex: Resource sheet TL 91, part TL 90.   J. D. Ambrose.
Report 73/8   ISBN 0 11 990614 9   £1.30

8   The sand and gravel resources of the country around Shotley and Felixstowe, Suffolk: Resource sheet TM 23.
R. Allender and S. E. Hollyer.
Report 73/13   ISBN 0 11 880625 4   £1.60

9   The sand and gravel resources of the country around Attlebridge, Norfolk: Resource sheet TG 11.   E. F. P. Nickless.
Report 73/15   ISBN 0 11 880658 0   £1.85

10   The sand and gravel resources of the country west of Colchester, Essex: Resource sheet TL 92.   J. D. Ambrose.
Report 74/6   ISBN 0 11 880671 8   £1.45

11   The sand and gravel resources of the country around Tattingstone, Suffolk: Resource sheet TM 13.   S. E. Hollyer.
Report 74/9   ISBN 0 11 880675 0   £1.95

12   The sand and gravel resources of the country around Gerrards Cross, Buckinghamshire: Resource sheet SU 99, TQ 08, 09.   H. C. Squirrell.
Report 74/14   ISBN 0 11 880710 2   £2.20

Mineral Assessment Reports

13   The sand and gravel resources of the country east of Chelmsford, Essex: Resource sheet TL 70.   M. R. Clarke.
ISBN 0 11 880744 7   £3.50

14   The sand and gravel resources of the country east of Colchester, Essex: Resource sheet TM 02.   J. D. Ambrose.
ISBN 0 11 880745 5   £3.25

15   The sand and gravel resources of the country around Newton on Trent, Lincolnshire: Resource sheet SK 87.
D. Price.
ISBN 0 11 880746 3   £3.00

16   The sand and gravel resources of the country around Braintree, Essex: Resource sheet TL 72.   M. R. Clarke.
ISBN 0 11 880747 1   £3.50

17   The sand and gravel resources of the country around Besthorpe, Nottinghamshire: Resource sheet SK 86, part SK 76.   J. R. Gozzard.
ISBN 0 11 880748 X   £3.00

18   The sand and gravel resources of the Thames Valley, the country around Cricklade, Wiltshire: Resource sheet SU 09 19, parts SP 00, 10.   P. R. Robson.
ISBN 0 11 880749 8   £3.00

19   The sand and gravel resources of the country south of Gainsborough, Lincolnshire: Resource sheet SK 88, part SK 78.   J. H. Lovell.
ISBN 0 11 880750 1   £2.50

20   The sand and gravel resources of the country east of Newark-upon-Trent, Nottinghamshire: Resource sheet SK 85.   J. R. Gozzard.
ISBN 0 11 880751 X   £2.75

21   The sand and gravel resources of the Thames and Kennet Valleys, the country around Pangbourne, Berkshire: Resource sheet SU 67.   H. C. Squirrell.
ISBN 0 11 880752 8   £3.25

22   The sand and gravel resources of the country north-west of Scunthorpe, Humberside: Resource sheet SE 81.
J. W. C. James.
ISBN 0 11 880753 6   £3.00

23   The sand and gravel resources of the Thames Valley, the country between Lechlade and Standlake: Resource sheet SP 30, parts SP 20, SU 29, 39.   P. Robson.
ISBN 0 11 881252 1   £7.25

24   The sand and gravel resources of the country around Aldermaston, Berkshire: Parts of resource sheet SU 56, 66.
H. C. Squirrell.
ISBN 0 11 881253 X   £5.00

25   The celestite resources of the area north-east of Bristol: Resource sheet ST 68, parts ST 59, 69, 79, 58, 78, 68, 77.
E. F. P. Nickless, S. J. Booth and P. N. Mosley.
ISBN 0 11 881262 9   £5.00

26   The limestone and dolomite resources of the country around Monyash, Derbyshire: Resource sheet SK 16.
F. C. Cox and D. McC. Bridge.
ISBN 0 11 881263 7   £7.00

27   The sand and gravel resources of the country west and south of Lincoln, Lincolnshire: Resource sheet SK 95, 96, SK 97.   I. Jackson.
ISBN 0 11 884003 7   £6.00

28   The sand and gravel resources of the country around Eynsham, Oxfordshire: Resource sheet SP 40, part SP 41.
W. J. R. Harries.
ISBN 0 11 884012 6   £3.00

29   The sand and gravel resources of the country south-west of Scunthorpe, Humberside: Resource sheet SE 80.
J. H. Lovell.
ISBN 0 11 884013 4   £3.50

30   Procedure for the assessment of limestone resources.
F. C. Cox, D. McC. Bridge and J. H. Hull.
ISBN 0 11 884030 4   £1.25

31   The sand and gravel resources of the country west of Newark-upon-Trent, Nottinghamshire. Resource sheet SK 75.   D. Price and P. J. Rogers.
ISBN 0 11 884031 2   £3.50

32   The sand and gravel resources of the country around Sonning and Henley: Resource sheet SU 77, 78.
H. C. Squirrell.
ISBN 0 11 884032 0   £5.25

33   The sand and gravel resources of the country north of Gainsborough: Resource sheet SK 89. J. Gozzard and D. Price.
ISBN 0 11 884033 9   £4.50

34   The sand and gravel resources of the Dengie Peninsula, Essex: Resource sheet TL 90, etc.   M. B. Simmons.
ISBN 0 11 884081 9   £5.00

35   The sand and gravel resources of the country around
Darvel: Resource sheet NS 53, 63, etc.   E. F. P. Nickless
and others.
ISBN 0 11 884082 7   £7.00

36   The sand and gravel resources of the country around
Southend-on-Sea, Essex: Resource sheets TQ 78, 79 etc.
S. E. Hollyer and M. B. Simmons.
ISBN 0 11 884083 5   £7.50

37   The sand and gravel resources of the country around
Bawtry, South Yorkshire: Resource sheet SK 49.
A. R. Clayton.
ISBN 0 11 884053 3   £5.75

38   The sand and gravel resources of the country around
Abingdon, Oxfordshire: Resource sheet SU 49, 59, SP 40,
50.   C. E. Corser.
ISBN 0 11 884084 5   £5.50

39   The sand and gravel resources of the Blackwater
Valley (Aldershot) area: Resource sheet SU 85, 86, parts
S U 84, 94, 95, 96.   M. R. Clarke, A. J. Dixon.
ISBN 0 11 884085 1   *not yet priced*

40   The sand and gravel resources of the country west of
Darlington, County Durham: Resource sheet NZ 11, 21.
A. Smith.
ISBN 0 11 884086 X   *not yet priced*

41   The sand and gravel resources of the country around
Garmouth, Grampian Region: Resource sheet NJ 36.
E. F. P. Nickless.
ISBN 0 11 884090 8   *not yet priced*

42   The sand and gravel resources of the country around
Maidenhead and Marlow: Resource sheet SU 88, parts
SU 87, 97, 98.   P. N. Dunkley.
ISBN 0 11 884091 6   £5.00

Reports of the Institute of Geological Sciences

Other Reports
69/9   Sand and gravel resources of the inner Moray Firth.
A. L. Harris and J. D. Peacock.
ISBN 0 11 880106 6   35p

70/4   Sands and gravels of the southern counties of
Scotland.   G. A. Goodlet.
ISBN 0 11 880105 8   90p

72/8   The use and resources of moulding sand in Northern
Ireland.   R. A. Old.
ISBN 0 11 881594 0   30p

73/9   The superficial deposits of the Firth of Clyde and its
sea lochs.   C. E. Deegan, R. Kirby, I. Rae and R. Floyd.
ISBN 0 11 880617 3   95p

77/1   Sources of aggregate in Northern Ireland (2nd
edition).   I. B. Cameron.
ISBN 0 11 881279 3   70p

77/2   Sand and gravel resources of the Grampian Region.
J. D. Peacock and others.
ISBN 0 11 881282 3   80p

77/5   Sand and gravel resources of the Fife Region.
M. A. E. Browne.
ISBN 0 11 884004 5   60p

77/6   Sand and gravel resources of the Tayside Region.
I. B. Paterson.
ISBN 0 11 884008 8   £1.40

77/8   Sand and gravel resources of the Strathclyde Region.
I. B. Cameron and others.
ISBN 0 11 884028 2   £2.50

77/9   Sand and gravel resources of the Central Region.
Scotland.   M. A. E. Browne.
ISBN 0 11 884016 9   £1.35

77/19   Sand and gravel resources of the Borders Region,
Scotland.   A. D. McAdam.
ISBN 0 11 884025 8   £1.00

77/22   Sand and gravel resources of the Dumfries and
Galloway Region of Scotland.   I. B. Cameron.
ISBN 0 11 884025 8   £1.20

78/1   Sand and gravels of the Lothian Region of Scotland.
A. D. McAdam.
ISBN 0 11 884042 8   £1.00

Dd 595773   K8

Typeset for the Institute of Geological Sciences
by Frowde and Co. (Printers) Limited, London SE5

Printed in England for Her Majesty's Stationery
Office by Commercial Colour Press, London E7